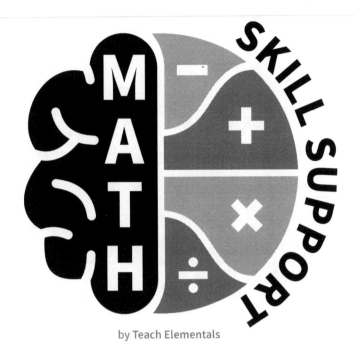

by Teach Elementals

Multiplication Facts 0-12

ISBN: 978-1-964695-00-6

© 2024 Teach Elementals, LLC. All rights reserved.

This work is reproducible by the original owner for personal use within their home or use with their individual classroom's students. No part of this publication may be distributed in any form or by any means, or stored in a database or retrieval system, without prior written consent of Teach Elementals, LLC, including, but not limited to, in any network or other electronic storage or transmission, or broadcast for distance learning.

Contents

Section A: Multiplication Facts by Number Families

Section B: Multiplication Facts Mixed Review

Find more of our products at
www.teachelementals.com

Name: _____ Date: _____

SCORE

1.
$$\begin{array}{r} 12 \\ \times\ 0 \\ \hline \end{array}$$

2.
$$\begin{array}{r} 0 \\ \times\ 9 \\ \hline \end{array}$$

3.
$$\begin{array}{r} 0 \\ \times\ 8 \\ \hline \end{array}$$

4.
$$\begin{array}{r} 0 \\ \times 10 \\ \hline \end{array}$$

5.
$$\begin{array}{r} 11 \\ \times\ 0 \\ \hline \end{array}$$

6.
$$\begin{array}{r} 3 \\ \times\ 0 \\ \hline \end{array}$$

7.
$$\begin{array}{r} 0 \\ \times\ 0 \\ \hline \end{array}$$

8.
$$\begin{array}{r} 0 \\ \times\ 6 \\ \hline \end{array}$$

9.
$$\begin{array}{r} 8 \\ \times\ 0 \\ \hline \end{array}$$

10.
$$\begin{array}{r} 0 \\ \times 12 \\ \hline \end{array}$$

11.
$$\begin{array}{r} 1 \\ \times\ 0 \\ \hline \end{array}$$

12.
$$\begin{array}{r} 10 \\ \times\ 0 \\ \hline \end{array}$$

13.
$$\begin{array}{r} 0 \\ \times\ 7 \\ \hline \end{array}$$

14.
$$\begin{array}{r} 0 \\ \times\ 5 \\ \hline \end{array}$$

15.
$$\begin{array}{r} 0 \\ \times\ 4 \\ \hline \end{array}$$

16.
$$\begin{array}{r} 0 \\ \times\ 3 \\ \hline \end{array}$$

17.
$$\begin{array}{r} 9 \\ \times\ 0 \\ \hline \end{array}$$

18.
$$\begin{array}{r} 2 \\ \times\ 0 \\ \hline \end{array}$$

19.
$$\begin{array}{r} 0 \\ \times 11 \\ \hline \end{array}$$

20.
$$\begin{array}{r} 0 \\ \times\ 2 \\ \hline \end{array}$$

21.
$$\begin{array}{r} 0 \\ \times\ 1 \\ \hline \end{array}$$

22.
$$\begin{array}{r} 5 \\ \times\ 0 \\ \hline \end{array}$$

23.
$$\begin{array}{r} 4 \\ \times\ 0 \\ \hline \end{array}$$

24.
$$\begin{array}{r} 6 \\ \times\ 0 \\ \hline \end{array}$$

25.
$$\begin{array}{r} 7 \\ \times\ 0 \\ \hline \end{array}$$

26.
$$\begin{array}{r} 0 \\ \times\ 5 \\ \hline \end{array}$$

27.
$$\begin{array}{r} 0 \\ \times\ 0 \\ \hline \end{array}$$

28.
$$\begin{array}{r} 8 \\ \times\ 0 \\ \hline \end{array}$$

29.
$$\begin{array}{r} 0 \\ \times\ 3 \\ \hline \end{array}$$

30.
$$\begin{array}{r} 0 \\ \times\ 4 \\ \hline \end{array}$$

31.
$$\begin{array}{r} 0 \\ \times\ 6 \\ \hline \end{array}$$

32.
$$\begin{array}{r} 2 \\ \times\ 0 \\ \hline \end{array}$$

33.
$$\begin{array}{r} 0 \\ \times\ 1 \\ \hline \end{array}$$

34.
$$\begin{array}{r} 11 \\ \times\ 0 \\ \hline \end{array}$$

35.
$$\begin{array}{r} 10 \\ \times\ 0 \\ \hline \end{array}$$

36.
$$\begin{array}{r} 9 \\ \times\ 0 \\ \hline \end{array}$$

37.
$$\begin{array}{r} 0 \\ \times\ 2 \\ \hline \end{array}$$

38.
$$\begin{array}{r} 5 \\ \times\ 0 \\ \hline \end{array}$$

39.
$$\begin{array}{r} 6 \\ \times\ 0 \\ \hline \end{array}$$

40.
$$\begin{array}{r} 1 \\ \times\ 0 \\ \hline \end{array}$$

41.
$$\begin{array}{r} 0 \\ \times 12 \\ \hline \end{array}$$

42.
$$\begin{array}{r} 0 \\ \times 10 \\ \hline \end{array}$$

43.
$$\begin{array}{r} 4 \\ \times\ 0 \\ \hline \end{array}$$

44.
$$\begin{array}{r} 12 \\ \times\ 0 \\ \hline \end{array}$$

45.
$$\begin{array}{r} 0 \\ \times\ 9 \\ \hline \end{array}$$

46.
$$\begin{array}{r} 0 \\ \times\ 8 \\ \hline \end{array}$$

47.
$$\begin{array}{r} 3 \\ \times\ 0 \\ \hline \end{array}$$

48.
$$\begin{array}{r} 7 \\ \times\ 0 \\ \hline \end{array}$$

49.
$$\begin{array}{r} 0 \\ \times 11 \\ \hline \end{array}$$

50.
$$\begin{array}{r} 0 \\ \times\ 7 \\ \hline \end{array}$$

51.
$$\begin{array}{r} 0 \\ \times\ 4 \\ \hline \end{array}$$

52.
$$\begin{array}{r} 0 \\ \times\ 5 \\ \hline \end{array}$$

53.
$$\begin{array}{r} 5 \\ \times\ 0 \\ \hline \end{array}$$

54.
$$\begin{array}{r} 0 \\ \times\ 7 \\ \hline \end{array}$$

55.
$$\begin{array}{r} 0 \\ \times\ 0 \\ \hline \end{array}$$

56.
$$\begin{array}{r} 8 \\ \times\ 0 \\ \hline \end{array}$$

57.
$$\begin{array}{r} 10 \\ \times\ 0 \\ \hline \end{array}$$

58.
$$\begin{array}{r} 7 \\ \times\ 0 \\ \hline \end{array}$$

59.
$$\begin{array}{r} 0 \\ \times\ 6 \\ \hline \end{array}$$

60.
$$\begin{array}{r} 6 \\ \times\ 0 \\ \hline \end{array}$$

© 2024 Teach Elementals, LLC

Name: _____   Date: _____

SCORE

1	**2**	**3**	**4**
0 × 2	0 × 7	0 × 6	8 × 0
5	**6**	**7**	**8**
3 × 0	0 × 4	0 × 10	0 × 5
9	**10**	**11**	**12**
0 × 8	0 × 12	5 × 0	2 × 0
13	**14**	**15**	**16**
0 × 11	12 × 0	0 × 1	10 × 0
17	**18**	**19**	**20**
11 × 0	0 × 0	0 × 3	1 × 0
21	**22**	**23**	**24**
7 × 0	6 × 0	4 × 0	9 × 0
25	**26**	**27**	**28**
0 × 9	0 × 7	8 × 0	0 × 1
29	**30**	**31**	**32**
0 × 5	0 × 8	12 × 0	2 × 0
33	**34**	**35**	**36**
4 × 0	9 × 0	7 × 0	0 × 4
37	**38**	**39**	**40**
3 × 0	0 × 11	0 × 6	0 × 10
41	**42**	**43**	**44**
0 × 0	10 × 0	6 × 0	0 × 2
45	**46**	**47**	**48**
0 × 9	1 × 0	0 × 3	0 × 12
49	**50**	**51**	**52**
11 × 0	5 × 0	1 × 0	6 × 0
53	**54**	**55**	**56**
5 × 0	12 × 0	0 × 3	0 × 5
57	**58**	**59**	**60**
0 × 7	0 × 0	0 × 9	0 × 12

© 2024 Teach Elementals, LLC

Name: _____

Date: _____

SCORE

1 $\begin{array}{r} 1 \\ \times 11 \\ \hline \end{array}$	**2** $\begin{array}{r} 6 \\ \times\ 1 \\ \hline \end{array}$	**3** $\begin{array}{r} 1 \\ \times 12 \\ \hline \end{array}$	**4** $\begin{array}{r} 7 \\ \times\ 1 \\ \hline \end{array}$

5 $\begin{array}{r} 1 \\ \times\ 3 \\ \hline \end{array}$	**6** $\begin{array}{r} 1 \\ \times\ 9 \\ \hline \end{array}$	**7** $\begin{array}{r} 1 \\ \times 10 \\ \hline \end{array}$	**8** $\begin{array}{r} 11 \\ \times\ 1 \\ \hline \end{array}$	**9** $\begin{array}{r} 4 \\ \times\ 1 \\ \hline \end{array}$	**10** $\begin{array}{r} 1 \\ \times\ 5 \\ \hline \end{array}$	**11** $\begin{array}{r} 0 \\ \times\ 1 \\ \hline \end{array}$	**12** $\begin{array}{r} 1 \\ \times\ 1 \\ \hline \end{array}$
13 $\begin{array}{r} 12 \\ \times\ 1 \\ \hline \end{array}$	**14** $\begin{array}{r} 10 \\ \times\ 1 \\ \hline \end{array}$	**15** $\begin{array}{r} 1 \\ \times\ 7 \\ \hline \end{array}$	**16** $\begin{array}{r} 1 \\ \times\ 2 \\ \hline \end{array}$	**17** $\begin{array}{r} 8 \\ \times\ 1 \\ \hline \end{array}$	**18** $\begin{array}{r} 9 \\ \times\ 1 \\ \hline \end{array}$	**19** $\begin{array}{r} 1 \\ \times\ 8 \\ \hline \end{array}$	**20** $\begin{array}{r} 1 \\ \times\ 0 \\ \hline \end{array}$
21 $\begin{array}{r} 1 \\ \times\ 4 \\ \hline \end{array}$	**22** $\begin{array}{r} 2 \\ \times\ 1 \\ \hline \end{array}$	**23** $\begin{array}{r} 5 \\ \times\ 1 \\ \hline \end{array}$	**24** $\begin{array}{r} 1 \\ \times\ 6 \\ \hline \end{array}$	**25** $\begin{array}{r} 3 \\ \times\ 1 \\ \hline \end{array}$	**26** $\begin{array}{r} 12 \\ \times\ 1 \\ \hline \end{array}$	**27** $\begin{array}{r} 1 \\ \times\ 8 \\ \hline \end{array}$	**28** $\begin{array}{r} 3 \\ \times\ 1 \\ \hline \end{array}$
29 $\begin{array}{r} 1 \\ \times\ 6 \\ \hline \end{array}$	**30** $\begin{array}{r} 1 \\ \times\ 5 \\ \hline \end{array}$	**31** $\begin{array}{r} 2 \\ \times\ 1 \\ \hline \end{array}$	**32** $\begin{array}{r} 1 \\ \times\ 1 \\ \hline \end{array}$	**33** $\begin{array}{r} 1 \\ \times 12 \\ \hline \end{array}$	**34** $\begin{array}{r} 1 \\ \times\ 2 \\ \hline \end{array}$	**35** $\begin{array}{r} 5 \\ \times\ 1 \\ \hline \end{array}$	**36** $\begin{array}{r} 1 \\ \times\ 3 \\ \hline \end{array}$
37 $\begin{array}{r} 1 \\ \times 11 \\ \hline \end{array}$	**38** $\begin{array}{r} 1 \\ \times 10 \\ \hline \end{array}$	**39** $\begin{array}{r} 4 \\ \times\ 1 \\ \hline \end{array}$	**40** $\begin{array}{r} 11 \\ \times\ 1 \\ \hline \end{array}$	**41** $\begin{array}{r} 1 \\ \times\ 0 \\ \hline \end{array}$	**42** $\begin{array}{r} 0 \\ \times\ 1 \\ \hline \end{array}$	**43** $\begin{array}{r} 1 \\ \times\ 4 \\ \hline \end{array}$	**44** $\begin{array}{r} 7 \\ \times\ 1 \\ \hline \end{array}$
45 $\begin{array}{r} 9 \\ \times\ 1 \\ \hline \end{array}$	**46** $\begin{array}{r} 1 \\ \times\ 9 \\ \hline \end{array}$	**47** $\begin{array}{r} 6 \\ \times\ 1 \\ \hline \end{array}$	**48** $\begin{array}{r} 8 \\ \times\ 1 \\ \hline \end{array}$	**49** $\begin{array}{r} 1 \\ \times\ 7 \\ \hline \end{array}$	**50** $\begin{array}{r} 10 \\ \times\ 1 \\ \hline \end{array}$	**51** $\begin{array}{r} 0 \\ \times\ 1 \\ \hline \end{array}$	**52** $\begin{array}{r} 1 \\ \times\ 1 \\ \hline \end{array}$
53 $\begin{array}{r} 3 \\ \times\ 1 \\ \hline \end{array}$	**54** $\begin{array}{r} 4 \\ \times\ 1 \\ \hline \end{array}$	**55** $\begin{array}{r} 6 \\ \times\ 1 \\ \hline \end{array}$	**56** $\begin{array}{r} 1 \\ \times\ 9 \\ \hline \end{array}$	**57** $\begin{array}{r} 11 \\ \times\ 1 \\ \hline \end{array}$	**58** $\begin{array}{r} 9 \\ \times\ 1 \\ \hline \end{array}$	**59** $\begin{array}{r} 1 \\ \times\ 7 \\ \hline \end{array}$	**60** $\begin{array}{r} 1 \\ \times\ 6 \\ \hline \end{array}$

© 2024 Teach Elementals, LLC

MULTIPLYING BY 1, SET B

Name: _____   Date: _____

SCORE

1
1
× 5

2
6
× 1

3
2
× 1

4
1
× 9

5
1
× 2

6
5
× 1

7
4
× 1

8
1
× 7

9
7
× 1

10
1
× 6

11
1
×11

12
0
× 1

13
3
× 1

14
12
× 1

15
1
×12

16
11
× 1

17
1
× 4

18
1
× 8

19
1
×10

20
1
× 0

21
8
× 1

22
9
× 1

23
1
× 1

24
10
× 1

25
1
× 3

26
1
×12

27
1
× 8

28
3
× 1

29
7
× 1

30
4
× 1

31
1
×10

32
1
× 7

33
6
× 1

34
1
× 4

35
1
×11

36
1
× 6

37
12
× 1

38
1
× 0

39
11
× 1

40
8
× 1

41
10
× 1

42
1
× 5

43
9
× 1

44
1
× 2

45
5
× 1

46
1
× 3

47
1
× 9

48
2
× 1

49
0
× 1

50
1
× 1

51
1
×10

52
4
× 1

53
1
× 5

54
10
× 1

55
2
× 1

56
1
×11

57
0
× 1

58
1
× 4

59
11
× 1

60
1
× 0

© 2024 Teach Elementals, LLC

Name: _____   Date: _____

SCORE

1
  5
× 2

2
  10
× 2

3
  2
×12

4
  2
× 7

5
  3
× 2

6
  12
× 2

7
  8
× 2

8
  2
× 9

9
  2
×11

10
  2
× 4

11
  7
× 2

12
  6
× 2

13
  2
× 0

14
  9
× 2

15
  2
× 2

16
  1
× 2

17
  4
× 2

18
  0
× 2

19
  11
× 2

20
  2
× 6

21
  2
× 3

22
  2
×10

23
  2
× 5

24
  2
× 1

25
  2
× 8

26
  10
× 2

27
  1
× 2

28
  5
× 2

29
  0
× 2

30
  2
× 0

31
  2
×12

32
  8
× 2

33
  2
× 8

34
  2
× 5

35
  3
× 2

36
  2
× 2

37
  2
× 7

38
  2
× 6

39
  2
× 3

40
  2
× 4

41
  7
× 2

42
  4
× 2

43
  9
× 2

44
  6
× 2

45
  2
×10

46
  2
× 9

47
  2
×11

48
  2
× 1

49
  12
× 2

50
  11
× 2

51
  10
× 2

52
  6
× 2

53
  0
× 2

54
  2
× 6

55
  9
× 2

56
  2
×10

57
  2
× 2

58
  3
× 2

59
  2
× 9

60
  2
× 7

© 2024 Teach Elementals, LLC

MULTIPLYING BY 2, SET B

Name: _____

Date: _____

SCORE

1 3 × 2	**2** 10 × 2	**3** 9 × 2	**4** 2 × 8
5 2 ×10	**6** 7 × 2	**7** 5 × 2	**8** 2 ×12
9 2 × 7	**10** 2 × 4	**11** 4 × 2	**12** 6 × 2
13 2 × 9	**14** 2 ×11	**15** 11 × 2	**16** 2 × 2
17 1 × 2	**18** 8 × 2	**19** 2 × 0	**20** 2 × 1
21 2 × 5	**22** 12 × 2	**23** 0 × 2	**24** 2 × 3
25 2 × 6	**26** 9 × 2	**27** 2 ×11	**28** 0 × 2
29 2 × 2	**30** 10 × 2	**31** 2 × 7	**32** 12 × 2
33 2 × 5	**34** 2 ×10	**35** 2 × 1	**36** 7 × 2
37 2 × 4	**38** 11 × 2	**39** 2 × 3	**40** 3 × 2
41 4 × 2	**42** 2 × 6	**43** 1 × 2	**44** 8 × 2
45 2 × 9	**46** 5 × 2	**47** 2 × 0	**48** 2 × 8
49 6 × 2	**50** 2 ×12	**51** 12 × 2	**52** 6 × 2
53 4 × 2	**54** 2 × 1	**55** 1 × 2	**56** 2 × 2
57 0 × 2	**58** 2 ×11	**59** 7 × 2	**60** 2 × 0

© 2024 Teach Elementals, LLC

Name: _____   Date: _____

SCORE

1 0 × 3	**2** 3 × 8	**3** 4 × 3	**4** 3 × 1
5 3 × 7	**6** 2 × 3	**7** 3 × 9	**8** 3 × 5
9 3 × 6	**10** 6 × 3	**11** 3 × 2	**12** 5 × 3
13 3 × 0	**14** 11 × 3	**15** 3 ×11	**16** 9 × 3
17 8 × 3	**18** 3 ×10	**19** 3 × 3	**20** 3 ×12
21 3 × 4	**22** 1 × 3	**23** 12 × 3	**24** 7 × 3
25 10 × 3	**26** 3 × 3	**27** 3 ×10	**28** 3 × 6
29 10 × 3	**30** 3 × 5	**31** 3 × 2	**32** 8 × 3
33 12 × 3	**34** 3 × 9	**35** 0 × 3	**36** 3 ×11
37 1 × 3	**38** 3 × 1	**39** 6 × 3	**40** 11 × 3
41 5 × 3	**42** 3 × 4	**43** 3 × 7	**44** 3 ×12
45 4 × 3	**46** 7 × 3	**47** 9 × 3	**48** 3 × 8
49 2 × 3	**50** 3 × 0	**51** 9 × 3	**52** 3 × 7
53 3 × 5	**54** 7 × 3	**55** 3 × 0	**56** 5 × 3
57 3 ×11	**58** 4 × 3	**59** 3 × 4	**60** 3 × 3

© 2024 Teach Elementals, LLC

SKILL SUPPORT

Name: _____

Date: _____

SCORE

1 $\begin{array}{r} 3 \\ \times\ 5 \\ \hline \end{array}$	**2** $\begin{array}{r} 3 \\ \times\ 2 \\ \hline \end{array}$	**3** $\begin{array}{r} 8 \\ \times\ 3 \\ \hline \end{array}$	**4** $\begin{array}{r} 7 \\ \times\ 3 \\ \hline \end{array}$

5 $\begin{array}{r} 12 \\ \times\ 3 \\ \hline \end{array}$	**6** $\begin{array}{r} 1 \\ \times\ 3 \\ \hline \end{array}$	**7** $\begin{array}{r} 4 \\ \times\ 3 \\ \hline \end{array}$	**8** $\begin{array}{r} 9 \\ \times\ 3 \\ \hline \end{array}$	**9** $\begin{array}{r} 3 \\ \times 11 \\ \hline \end{array}$	**10** $\begin{array}{r} 6 \\ \times\ 3 \\ \hline \end{array}$	**11** $\begin{array}{r} 5 \\ \times\ 3 \\ \hline \end{array}$	**12** $\begin{array}{r} 2 \\ \times\ 3 \\ \hline \end{array}$
13 $\begin{array}{r} 3 \\ \times\ 1 \\ \hline \end{array}$	**14** $\begin{array}{r} 10 \\ \times\ 3 \\ \hline \end{array}$	**15** $\begin{array}{r} 3 \\ \times\ 9 \\ \hline \end{array}$	**16** $\begin{array}{r} 3 \\ \times\ 7 \\ \hline \end{array}$	**17** $\begin{array}{r} 3 \\ \times\ 6 \\ \hline \end{array}$	**18** $\begin{array}{r} 3 \\ \times\ 8 \\ \hline \end{array}$	**19** $\begin{array}{r} 3 \\ \times 10 \\ \hline \end{array}$	**20** $\begin{array}{r} 3 \\ \times\ 0 \\ \hline \end{array}$
21 $\begin{array}{r} 3 \\ \times 12 \\ \hline \end{array}$	**22** $\begin{array}{r} 3 \\ \times\ 4 \\ \hline \end{array}$	**23** $\begin{array}{r} 11 \\ \times\ 3 \\ \hline \end{array}$	**24** $\begin{array}{r} 0 \\ \times\ 3 \\ \hline \end{array}$	**25** $\begin{array}{r} 3 \\ \times\ 3 \\ \hline \end{array}$	**26** $\begin{array}{r} 11 \\ \times\ 3 \\ \hline \end{array}$	**27** $\begin{array}{r} 3 \\ \times\ 3 \\ \hline \end{array}$	**28** $\begin{array}{r} 0 \\ \times\ 3 \\ \hline \end{array}$
29 $\begin{array}{r} 10 \\ \times\ 3 \\ \hline \end{array}$	**30** $\begin{array}{r} 3 \\ \times\ 1 \\ \hline \end{array}$	**31** $\begin{array}{r} 9 \\ \times\ 3 \\ \hline \end{array}$	**32** $\begin{array}{r} 5 \\ \times\ 3 \\ \hline \end{array}$	**33** $\begin{array}{r} 3 \\ \times\ 7 \\ \hline \end{array}$	**34** $\begin{array}{r} 3 \\ \times\ 6 \\ \hline \end{array}$	**35** $\begin{array}{r} 3 \\ \times\ 8 \\ \hline \end{array}$	**36** $\begin{array}{r} 8 \\ \times\ 3 \\ \hline \end{array}$
37 $\begin{array}{r} 12 \\ \times\ 3 \\ \hline \end{array}$	**38** $\begin{array}{r} 1 \\ \times\ 3 \\ \hline \end{array}$	**39** $\begin{array}{r} 3 \\ \times\ 5 \\ \hline \end{array}$	**40** $\begin{array}{r} 3 \\ \times\ 9 \\ \hline \end{array}$	**41** $\begin{array}{r} 2 \\ \times\ 3 \\ \hline \end{array}$	**42** $\begin{array}{r} 3 \\ \times\ 4 \\ \hline \end{array}$	**43** $\begin{array}{r} 3 \\ \times\ 0 \\ \hline \end{array}$	**44** $\begin{array}{r} 4 \\ \times\ 3 \\ \hline \end{array}$
45 $\begin{array}{r} 3 \\ \times\ 2 \\ \hline \end{array}$	**46** $\begin{array}{r} 3 \\ \times 11 \\ \hline \end{array}$	**47** $\begin{array}{r} 7 \\ \times\ 3 \\ \hline \end{array}$	**48** $\begin{array}{r} 6 \\ \times\ 3 \\ \hline \end{array}$	**49** $\begin{array}{r} 3 \\ \times 10 \\ \hline \end{array}$	**50** $\begin{array}{r} 3 \\ \times 12 \\ \hline \end{array}$	**51** $\begin{array}{r} 1 \\ \times\ 3 \\ \hline \end{array}$	**52** $\begin{array}{r} 5 \\ \times\ 3 \\ \hline \end{array}$
53 $\begin{array}{r} 3 \\ \times\ 8 \\ \hline \end{array}$	**54** $\begin{array}{r} 3 \\ \times\ 7 \\ \hline \end{array}$	**55** $\begin{array}{r} 3 \\ \times 11 \\ \hline \end{array}$	**56** $\begin{array}{r} 2 \\ \times\ 3 \\ \hline \end{array}$	**57** $\begin{array}{r} 9 \\ \times\ 3 \\ \hline \end{array}$	**58** $\begin{array}{r} 3 \\ \times 12 \\ \hline \end{array}$	**59** $\begin{array}{r} 3 \\ \times\ 5 \\ \hline \end{array}$	**60** $\begin{array}{r} 6 \\ \times\ 3 \\ \hline \end{array}$

© 2024 Teach Elementals, LLC

Name: _____   **Date:** _____

SCORE

1 $\begin{array}{r} 3 \\ \times\ 4 \\ \hline \end{array}$	**2** $\begin{array}{r} 11 \\ \times\ 3 \\ \hline \end{array}$	**3** $\begin{array}{r} 3 \\ \times\ 1 \\ \hline \end{array}$	**4** $\begin{array}{r} 9 \\ \times\ 3 \\ \hline \end{array}$

5 $\begin{array}{r} 1 \\ \times\ 3 \\ \hline \end{array}$	**6** $\begin{array}{r} 5 \\ \times\ 3 \\ \hline \end{array}$	**7** $\begin{array}{r} 3 \\ \times\ 9 \\ \hline \end{array}$	**8** $\begin{array}{r} 10 \\ \times\ 3 \\ \hline \end{array}$	**9** $\begin{array}{r} 3 \\ \times 12 \\ \hline \end{array}$	**10** $\begin{array}{r} 12 \\ \times\ 3 \\ \hline \end{array}$	**11** $\begin{array}{r} 3 \\ \times\ 7 \\ \hline \end{array}$	**12** $\begin{array}{r} 8 \\ \times\ 3 \\ \hline \end{array}$
13 $\begin{array}{r} 4 \\ \times\ 3 \\ \hline \end{array}$	**14** $\begin{array}{r} 6 \\ \times\ 3 \\ \hline \end{array}$	**15** $\begin{array}{r} 3 \\ \times\ 3 \\ \hline \end{array}$	**16** $\begin{array}{r} 0 \\ \times\ 3 \\ \hline \end{array}$	**17** $\begin{array}{r} 3 \\ \times 10 \\ \hline \end{array}$	**18** $\begin{array}{r} 2 \\ \times\ 3 \\ \hline \end{array}$	**19** $\begin{array}{r} 3 \\ \times\ 2 \\ \hline \end{array}$	**20** $\begin{array}{r} 3 \\ \times\ 8 \\ \hline \end{array}$
21 $\begin{array}{r} 3 \\ \times 11 \\ \hline \end{array}$	**22** $\begin{array}{r} 3 \\ \times\ 6 \\ \hline \end{array}$	**23** $\begin{array}{r} 3 \\ \times\ 5 \\ \hline \end{array}$	**24** $\begin{array}{r} 7 \\ \times\ 3 \\ \hline \end{array}$	**25** $\begin{array}{r} 3 \\ \times\ 0 \\ \hline \end{array}$	**26** $\begin{array}{r} 3 \\ \times\ 1 \\ \hline \end{array}$	**27** $\begin{array}{r} 11 \\ \times\ 3 \\ \hline \end{array}$	**28** $\begin{array}{r} 6 \\ \times\ 3 \\ \hline \end{array}$
29 $\begin{array}{r} 10 \\ \times\ 3 \\ \hline \end{array}$	**30** $\begin{array}{r} 3 \\ \times\ 0 \\ \hline \end{array}$	**31** $\begin{array}{r} 3 \\ \times 10 \\ \hline \end{array}$	**32** $\begin{array}{r} 3 \\ \times\ 6 \\ \hline \end{array}$	**33** $\begin{array}{r} 5 \\ \times\ 3 \\ \hline \end{array}$	**34** $\begin{array}{r} 3 \\ \times\ 9 \\ \hline \end{array}$	**35** $\begin{array}{r} 9 \\ \times\ 3 \\ \hline \end{array}$	**36** $\begin{array}{r} 3 \\ \times\ 5 \\ \hline \end{array}$
37 $\begin{array}{r} 0 \\ \times\ 3 \\ \hline \end{array}$	**38** $\begin{array}{r} 1 \\ \times\ 3 \\ \hline \end{array}$	**39** $\begin{array}{r} 3 \\ \times 12 \\ \hline \end{array}$	**40** $\begin{array}{r} 3 \\ \times 11 \\ \hline \end{array}$	**41** $\begin{array}{r} 3 \\ \times\ 7 \\ \hline \end{array}$	**42** $\begin{array}{r} 12 \\ \times\ 3 \\ \hline \end{array}$	**43** $\begin{array}{r} 3 \\ \times\ 4 \\ \hline \end{array}$	**44** $\begin{array}{r} 3 \\ \times\ 2 \\ \hline \end{array}$
45 $\begin{array}{r} 2 \\ \times\ 3 \\ \hline \end{array}$	**46** $\begin{array}{r} 7 \\ \times\ 3 \\ \hline \end{array}$	**47** $\begin{array}{r} 3 \\ \times\ 8 \\ \hline \end{array}$	**48** $\begin{array}{r} 4 \\ \times\ 3 \\ \hline \end{array}$	**49** $\begin{array}{r} 8 \\ \times\ 3 \\ \hline \end{array}$	**50** $\begin{array}{r} 3 \\ \times\ 3 \\ \hline \end{array}$	**51** $\begin{array}{r} 1 \\ \times\ 3 \\ \hline \end{array}$	**52** $\begin{array}{r} 3 \\ \times\ 1 \\ \hline \end{array}$
53 $\begin{array}{r} 3 \\ \times\ 3 \\ \hline \end{array}$	**54** $\begin{array}{r} 12 \\ \times\ 3 \\ \hline \end{array}$	**55** $\begin{array}{r} 0 \\ \times\ 3 \\ \hline \end{array}$	**56** $\begin{array}{r} 10 \\ \times\ 3 \\ \hline \end{array}$	**57** $\begin{array}{r} 3 \\ \times\ 4 \\ \hline \end{array}$	**58** $\begin{array}{r} 3 \\ \times\ 9 \\ \hline \end{array}$	**59** $\begin{array}{r} 7 \\ \times\ 3 \\ \hline \end{array}$	**60** $\begin{array}{r} 3 \\ \times\ 5 \\ \hline \end{array}$

© 2024 Teach Elementals, LLC

MULTIPLYING BY 4, SET A

Name: _____ Date: _____

SCORE

1
8
× 4

2
4
× 5

3
9
× 4

4
7
× 4

5
4
× 7

6
4
× 6

7
4
× 3

8
4
× 4

9
6
× 4

10
4
×11

11
10
× 4

12
4
× 0

13
3
× 4

14
2
× 4

15
4
× 9

16
12
× 4

17
4
×12

18
4
× 8

19
4
× 1

20
4
× 2

21
0
× 4

22
11
× 4

23
1
× 4

24
4
×10

25
5
× 4

26
9
× 4

27
4
× 0

28
10
× 4

29
4
× 4

30
4
× 3

31
0
× 4

32
12
× 4

33
4
× 9

34
6
× 4

35
8
× 4

36
3
× 4

37
4
×12

38
11
× 4

39
4
× 2

40
4
× 8

41
4
× 5

42
1
× 4

43
2
× 4

44
4
×10

45
4
× 1

46
4
×11

47
7
× 4

48
4
× 6

49
4
× 7

50
5
× 4

51
10
× 4

52
3
× 4

53
4
×11

54
4
× 8

55
0
× 4

56
8
× 4

57
7
× 4

58
4
× 6

59
4
× 9

60
4
× 2

© 2024 Teach Elementals, LLC

Name: _____ Date: _____

SCORE

1
```
    7
×   4
```

2
```
    4
× 11
```

3
```
   12
×   4
```

4
```
    4
×   4
```

5
```
   10
×   4
```

6
```
    4
×   3
```

7
```
    2
×   4
```

8
```
    4
×   0
```

9
```
    4
×   2
```

10
```
    4
×   7
```

11
```
    4
× 12
```

12
```
    6
×   4
```

13
```
    0
×   4
```

14
```
    4
× 10
```

15
```
    4
×   1
```

16
```
    4
×   6
```

17
```
    4
×   8
```

18
```
    5
×   4
```

19
```
   11
×   4
```

20
```
    1
×   4
```

21
```
    3
×   4
```

22
```
    4
×   9
```

23
```
    4
×   5
```

24
```
    9
×   4
```

25
```
    8
×   4
```

26
```
    4
×   9
```

27
```
    9
×   4
```

28
```
    8
×   4
```

29
```
    4
×   8
```

30
```
    6
×   4
```

31
```
    4
×   6
```

32
```
    4
×   7
```

33
```
    4
× 10
```

34
```
    3
×   4
```

35
```
    4
×   4
```

36
```
    2
×   4
```

37
```
    4
× 12
```

38
```
    4
× 11
```

39
```
    4
×   3
```

40
```
    4
×   5
```

41
```
    0
×   4
```

42
```
    4
×   0
```

43
```
    4
×   1
```

44
```
    5
×   4
```

45
```
    4
×   2
```

46
```
   10
×   4
```

47
```
   12
×   4
```

48
```
   11
×   4
```

49
```
    7
×   4
```

50
```
    1
×   4
```

51
```
    4
×   4
```

52
```
    7
×   4
```

53
```
    5
×   4
```

54
```
    4
×   2
```

55
```
    4
× 10
```

56
```
    2
×   4
```

57
```
    1
×   4
```

58
```
   11
×   4
```

59
```
    4
×   1
```

60
```
    8
×   4
```

© 2024 Teach Elementals, LLC

Skill Support: Multiplication Facts 0-12 11

MULTIPLYING BY 4, SET C

Name: _____

Date: _____

SCORE

#		#		#		#	
1	4 × 8	**2**	10 × 4	**3**	11 × 4	**4**	1 × 4

#		#		#		#		#		#		#		#	
5	4 × 3	**6**	4 × 1	**7**	4 × 0	**8**	4 × 5	**9**	4 × 9	**10**	6 × 4	**11**	12 × 4	**12**	4 × 7
13	7 × 4	**14**	5 × 4	**15**	4 ×11	**16**	3 × 4	**17**	2 × 4	**18**	4 × 2	**19**	0 × 4	**20**	4 ×10
21	9 × 4	**22**	8 × 4	**23**	4 × 6	**24**	4 ×12	**25**	4 × 4	**26**	4 × 9	**27**	6 × 4	**28**	4 ×11
29	3 × 4	**30**	4 ×12	**31**	4 × 3	**32**	4 × 1	**33**	4 ×10	**34**	9 × 4	**35**	4 × 6	**36**	4 × 7
37	0 × 4	**38**	1 × 4	**39**	11 × 4	**40**	2 × 4	**41**	7 × 4	**42**	4 × 4	**43**	8 × 4	**44**	5 × 4
45	12 × 4	**46**	4 × 5	**47**	4 × 0	**48**	4 × 8	**49**	10 × 4	**50**	4 × 2	**51**	4 × 1	**52**	1 × 4
53	10 × 4	**54**	4 × 2	**55**	4 × 5	**56**	4 × 8	**57**	4 × 0	**58**	4 × 3	**59**	4 × 9	**60**	4 ×11

© 2024 Teach Elementals, LLC

Name: _____   Date: _____

SCORE

1	**2**	**3**	**4**
0 ×5	9 ×5	10 ×5	5 ×6

5	**6**	**7**	**8**	**9**	**10**	**11**	**12**
3 ×5	5 ×9	5 ×2	8 ×5	5 ×8	2 ×5	5 ×1	7 ×5
13	**14**	**15**	**16**	**17**	**18**	**19**	**20**
5 ×12	5 ×11	5 ×0	5 ×3	4 ×5	12 ×5	11 ×5	6 ×5
21	**22**	**23**	**24**	**25**	**26**	**27**	**28**
5 ×4	1 ×5	5 ×7	5 ×10	5 ×5	9 ×5	5 ×12	5 ×3
29	**30**	**31**	**32**	**33**	**34**	**35**	**36**
7 ×5	11 ×5	5 ×9	5 ×8	3 ×5	5 ×10	5 ×0	5 ×4
37	**38**	**39**	**40**	**41**	**42**	**43**	**44**
0 ×5	5 ×6	1 ×5	5 ×2	4 ×5	5 ×5	10 ×5	5 ×1
45	**46**	**47**	**48**	**49**	**50**	**51**	**52**
8 ×5	5 ×7	2 ×5	6 ×5	12 ×5	5 ×11	4 ×5	5 ×12
53	**54**	**55**	**56**	**57**	**58**	**59**	**60**
7 ×5	5 ×10	5 ×1	12 ×5	5 ×11	5 ×8	5 ×3	2 ×5

© 2024 Teach Elementals, LLC

MULTIPLYING BY 5, SET B

DAY 14

Name: _____

Date: _____

SCORE

1 5 ×2	**2** 4 ×5	**3** 5 ×12	**4** 5 ×11

5 12 ×5	**6** 11 ×5	**7** 5 ×10	**8** 5 ×6	**9** 5 ×1	**10** 5 ×8	**11** 7 ×5	**12** 2 ×5
13 5 ×4	**14** 5 ×0	**15** 3 ×5	**16** 10 ×5	**17** 6 ×5	**18** 5 ×5	**19** 9 ×5	**20** 5 ×3
21 5 ×9	**22** 8 ×5	**23** 1 ×5	**24** 0 ×5	**25** 5 ×7	**26** 5 ×2	**27** 5 ×3	**28** 10 ×5
29 5 ×0	**30** 5 ×10	**31** 7 ×5	**32** 1 ×5	**33** 0 ×5	**34** 2 ×5	**35** 5 ×11	**36** 5 ×9
37 9 ×5	**38** 3 ×5	**39** 11 ×5	**40** 5 ×6	**41** 5 ×1	**42** 5 ×7	**43** 12 ×5	**44** 5 ×4
45 5 ×5	**46** 5 ×8	**47** 8 ×5	**48** 4 ×5	**49** 6 ×5	**50** 5 ×12	**51** 5 ×7	**52** 5 ×2
53 5 ×0	**54** 5 ×8	**55** 5 ×9	**56** 1 ×5	**57** 5 ×3	**58** 12 ×5	**59** 2 ×5	**60** 5 ×1

14   *Skill Support: Multiplication Facts 0-12*

© 2024 Teach Elementals, LLC

Name: _____ Date: _____

SCORE

1)
$$\begin{array}{r} 12 \\ \times\ 5 \\ \hline \end{array}$$

2)
$$\begin{array}{r} 5 \\ \times\ 8 \\ \hline \end{array}$$

3)
$$\begin{array}{r} 6 \\ \times\ 5 \\ \hline \end{array}$$

4)
$$\begin{array}{r} 5 \\ \times\ 2 \\ \hline \end{array}$$

5)
$$\begin{array}{r} 5 \\ \times\ 0 \\ \hline \end{array}$$

6)
$$\begin{array}{r} 11 \\ \times\ 5 \\ \hline \end{array}$$

7)
$$\begin{array}{r} 5 \\ \times\ 7 \\ \hline \end{array}$$

8)
$$\begin{array}{r} 3 \\ \times\ 5 \\ \hline \end{array}$$

9)
$$\begin{array}{r} 5 \\ \times\ 1 \\ \hline \end{array}$$

10)
$$\begin{array}{r} 8 \\ \times\ 5 \\ \hline \end{array}$$

11)
$$\begin{array}{r} 5 \\ \times\ 5 \\ \hline \end{array}$$

12)
$$\begin{array}{r} 7 \\ \times\ 5 \\ \hline \end{array}$$

13)
$$\begin{array}{r} 5 \\ \times 11 \\ \hline \end{array}$$

14)
$$\begin{array}{r} 4 \\ \times\ 5 \\ \hline \end{array}$$

15)
$$\begin{array}{r} 5 \\ \times\ 3 \\ \hline \end{array}$$

16)
$$\begin{array}{r} 5 \\ \times\ 6 \\ \hline \end{array}$$

17)
$$\begin{array}{r} 5 \\ \times\ 4 \\ \hline \end{array}$$

18)
$$\begin{array}{r} 2 \\ \times\ 5 \\ \hline \end{array}$$

19)
$$\begin{array}{r} 5 \\ \times 12 \\ \hline \end{array}$$

20)
$$\begin{array}{r} 9 \\ \times\ 5 \\ \hline \end{array}$$

21)
$$\begin{array}{r} 0 \\ \times\ 5 \\ \hline \end{array}$$

22)
$$\begin{array}{r} 5 \\ \times 10 \\ \hline \end{array}$$

23)
$$\begin{array}{r} 10 \\ \times\ 5 \\ \hline \end{array}$$

24)
$$\begin{array}{r} 5 \\ \times\ 9 \\ \hline \end{array}$$

25)
$$\begin{array}{r} 1 \\ \times\ 5 \\ \hline \end{array}$$

26)
$$\begin{array}{r} 5 \\ \times\ 1 \\ \hline \end{array}$$

27)
$$\begin{array}{r} 12 \\ \times\ 5 \\ \hline \end{array}$$

28)
$$\begin{array}{r} 5 \\ \times 10 \\ \hline \end{array}$$

29)
$$\begin{array}{r} 5 \\ \times\ 9 \\ \hline \end{array}$$

30)
$$\begin{array}{r} 11 \\ \times\ 5 \\ \hline \end{array}$$

31)
$$\begin{array}{r} 6 \\ \times\ 5 \\ \hline \end{array}$$

32)
$$\begin{array}{r} 5 \\ \times\ 8 \\ \hline \end{array}$$

33)
$$\begin{array}{r} 2 \\ \times\ 5 \\ \hline \end{array}$$

34)
$$\begin{array}{r} 5 \\ \times\ 6 \\ \hline \end{array}$$

35)
$$\begin{array}{r} 5 \\ \times 11 \\ \hline \end{array}$$

36)
$$\begin{array}{r} 5 \\ \times\ 3 \\ \hline \end{array}$$

37)
$$\begin{array}{r} 9 \\ \times\ 5 \\ \hline \end{array}$$

38)
$$\begin{array}{r} 0 \\ \times\ 5 \\ \hline \end{array}$$

39)
$$\begin{array}{r} 5 \\ \times\ 2 \\ \hline \end{array}$$

40)
$$\begin{array}{r} 7 \\ \times\ 5 \\ \hline \end{array}$$

41)
$$\begin{array}{r} 8 \\ \times\ 5 \\ \hline \end{array}$$

42)
$$\begin{array}{r} 5 \\ \times\ 0 \\ \hline \end{array}$$

43)
$$\begin{array}{r} 10 \\ \times\ 5 \\ \hline \end{array}$$

44)
$$\begin{array}{r} 3 \\ \times\ 5 \\ \hline \end{array}$$

45)
$$\begin{array}{r} 5 \\ \times 12 \\ \hline \end{array}$$

46)
$$\begin{array}{r} 5 \\ \times\ 7 \\ \hline \end{array}$$

47)
$$\begin{array}{r} 4 \\ \times\ 5 \\ \hline \end{array}$$

48)
$$\begin{array}{r} 5 \\ \times\ 4 \\ \hline \end{array}$$

49)
$$\begin{array}{r} 5 \\ \times\ 5 \\ \hline \end{array}$$

50)
$$\begin{array}{r} 1 \\ \times\ 5 \\ \hline \end{array}$$

51)
$$\begin{array}{r} 5 \\ \times\ 8 \\ \hline \end{array}$$

52)
$$\begin{array}{r} 8 \\ \times\ 5 \\ \hline \end{array}$$

53)
$$\begin{array}{r} 5 \\ \times\ 3 \\ \hline \end{array}$$

54)
$$\begin{array}{r} 5 \\ \times\ 6 \\ \hline \end{array}$$

55)
$$\begin{array}{r} 6 \\ \times\ 5 \\ \hline \end{array}$$

56)
$$\begin{array}{r} 3 \\ \times\ 5 \\ \hline \end{array}$$

57)
$$\begin{array}{r} 5 \\ \times\ 9 \\ \hline \end{array}$$

58)
$$\begin{array}{r} 9 \\ \times\ 5 \\ \hline \end{array}$$

59)
$$\begin{array}{r} 5 \\ \times 11 \\ \hline \end{array}$$

60)
$$\begin{array}{r} 0 \\ \times\ 5 \\ \hline \end{array}$$

© 2024 Teach Elementals, LLC

MULTIPLYING BY 6, SET A

Name: _____   Date: _____

SCORE

1.
4
× 6

2.
6
×10

3.
6
× 2

4.
8
× 6

5.
6
× 9

6.
6
× 8

7.
6
× 3

8.
6
×11

9.
3
× 6

10.
5
× 6

11.
6
× 5

12.
1
× 6

13.
2
× 6

14.
6
× 4

15.
6
×12

16.
10
× 6

17.
0
× 6

18.
11
× 6

19.
6
× 7

20.
6
× 1

21.
9
× 6

22.
7
× 6

23.
6
× 6

24.
6
× 0

25.
12
× 6

26.
0
× 6

27.
7
× 6

28.
6
× 2

29.
6
× 6

30.
6
×10

31.
9
× 6

32.
6
× 8

33.
8
× 6

34.
10
× 6

35.
6
× 9

36.
4
× 6

37.
6
× 3

38.
6
×11

39.
6
× 0

40.
6
× 1

41.
3
× 6

42.
11
× 6

43.
6
×12

44.
2
× 6

45.
1
× 6

46.
6
× 5

47.
6
× 7

48.
6
× 4

49.
5
× 6

50.
12
× 6

51.
6
×10

52.
6
× 9

53.
4
× 6

54.
7
× 6

55.
1
× 6

56.
10
× 6

57.
0
× 6

58.
6
× 2

59.
11
× 6

60.
6
× 8

© 2024 Teach Elementals, LLC

Name: _____   Date: _____

SCORE

1
11
× 6

2
12
× 6

3
9
× 6

4
6
× 2

5
6
× 1

6
6
× 9

7
6
×10

8
6
× 3

9
6
× 8

10
2
× 6

11
8
× 6

12
7
× 6

13
4
× 6

14
6
×12

15
6
× 6

16
3
× 6

17
6
×11

18
10
× 6

19
6
× 4

20
5
× 6

21
6
× 7

22
6
× 0

23
6
× 5

24
1
× 6

25
0
× 6

26
6
× 2

27
9
× 6

28
1
× 6

29
6
× 6

30
4
× 6

31
6
×10

32
8
× 6

33
6
× 7

34
10
× 6

35
11
× 6

36
12
× 6

37
7
× 6

38
6
×12

39
2
× 6

40
6
× 0

41
3
× 6

42
6
× 1

43
6
× 4

44
6
× 5

45
6
×11

46
5
× 6

47
6
× 8

48
6
× 9

49
6
× 3

50
6
× 2

51
6
×12

52
9
× 6

53
12
× 6

54
6
× 1

55
6
×10

56
6
× 6

57
10
× 6

58
6
× 8

59
6
× 4

60
11
× 6

Name: _____   Date: _____

SCORE

1
9
× 6

2
8
× 6

3
6
×10

4
6
× 0

5
6
× 6

6
6
× 3

7
3
× 6

8
7
× 6

9
6
× 1

10
5
× 6

11
0
× 6

12
11
× 6

13
6
× 7

14
6
× 4

15
6
×11

16
6
× 2

17
2
× 6

18
4
× 6

19
6
× 9

20
6
×12

21
1
× 6

22
10
× 6

23
6
× 5

24
6
× 8

25
12
× 6

26
6
× 0

27
3
× 6

28
7
× 6

29
0
× 6

30
6
× 4

31
6
×10

32
4
× 6

33
6
× 6

34
1
× 6

35
6
× 1

36
6
× 7

37
10
× 6

38
6
× 2

39
9
× 6

40
12
× 6

41
6
× 9

42
6
×11

43
6
×12

44
6
× 3

45
6
× 8

46
11
× 6

47
5
× 6

48
8
× 6

49
2
× 6

50
6
× 5

51
11
× 6

52
6
× 8

53
6
× 6

54
6
×11

55
6
× 1

56
1
× 6

57
6
× 5

58
8
× 6

59
4
× 6

60
6
×10

© 2024 Teach Elementals, LLC

Name: _____   Date: _____

SCORE

1 4 × 6	**2** 12 × 6	**3** 10 × 6	**4** 6 × 12

5 6 × 9	**6** 7 × 6	**7** 9 × 6	**8** 6 × 0	**9** 6 × 2	**10** 2 × 6	**11** 6 × 4	**12** 6 × 3

13 6 × 11	**14** 5 × 6	**15** 0 × 6	**16** 6 × 8	**17** 11 × 6	**18** 8 × 6	**19** 6 × 1	**20** 6 × 6

21 6 × 5	**22** 3 × 6	**23** 1 × 6	**24** 6 × 7	**25** 6 × 10	**26** 6 × 5	**27** 10 × 6	**28** 6 × 6

29 6 × 7	**30** 11 × 6	**31** 6 × 4	**32** 9 × 6	**33** 4 × 6	**34** 6 × 9	**35** 3 × 6	**36** 6 × 10

37 5 × 6	**38** 6 × 8	**39** 12 × 6	**40** 6 × 1	**41** 8 × 6	**42** 6 × 0	**43** 7 × 6	**44** 6 × 12

45 0 × 6	**46** 6 × 11	**47** 6 × 2	**48** 2 × 6	**49** 1 × 6	**50** 6 × 3	**51** 6 × 0	**52** 2 × 6

53 10 × 6	**54** 6 × 9	**55** 4 × 6	**56** 6 × 4	**57** 6 × 6	**58** 7 × 6	**59** 6 × 5	**60** 9 × 6

© 2024 Teach Elementals, LLC

Name: _____

Date: _____

SCORE

1	2	3	4
7 × 8	7 ×11	11 × 7	2 × 7

5	6	7	8	9	10	11	12
7 × 0	0 × 7	9 × 7	7 × 3	7 ×12	8 × 7	3 × 7	7 × 5

13	14	15	16	17	18	19	20
1 × 7	7 × 2	12 × 7	6 × 7	5 × 7	7 × 9	7 × 6	7 ×10

21	22	23	24	25	26	27	28
7 × 4	4 × 7	10 × 7	7 × 1	7 × 7	7 × 1	7 × 6	12 × 7

29	30	31	32	33	34	35	36
4 × 7	3 × 7	7 ×12	10 × 7	5 × 7	9 × 7	7 ×10	8 × 7

37	38	39	40	41	42	43	44
7 × 7	7 × 5	7 × 8	2 × 7	6 × 7	7 ×11	7 × 3	7 × 0

45	46	47	48	49	50	51	52
7 × 9	7 × 2	1 × 7	7 × 4	11 × 7	0 × 7	7 × 8	5 × 7

53	54	55	56	57	58	59	60
9 × 7	2 × 7	6 × 7	7 × 5	7 × 9	7 ×11	4 × 7	10 × 7

© 2024 Teach Elementals, LLC

DAY 20

Name: _____   Date: _____

SCORE

1
```
  11
×  7
```

2
```
   0
×  7
```

3
```
   7
× 11
```

4
```
   7
×  0
```

5
```
   7
×  6
```

6
```
   7
×  1
```

7
```
   7
×  3
```

8
```
   5
×  7
```

9
```
   3
×  7
```

10
```
   7
×  9
```

11
```
   2
×  7
```

12
```
   7
× 10
```

13
```
   7
× 12
```

14
```
  12
×  7
```

15
```
   7
×  2
```

16
```
   1
×  7
```

17
```
   4
×  7
```

18
```
   7
×  7
```

19
```
   9
×  7
```

20
```
   8
×  7
```

21
```
   7
×  5
```

22
```
  10
×  7
```

23
```
   7
×  8
```

24
```
   7
×  4
```

25
```
   6
×  7
```

26
```
   7
×  5
```

27
```
  10
×  7
```

28
```
   5
×  7
```

29
```
   7
× 10
```

30
```
   7
×  3
```

31
```
   7
×  4
```

32
```
   7
× 12
```

33
```
   7
×  2
```

34
```
   0
×  7
```

35
```
   7
×  7
```

36
```
  11
×  7
```

37
```
   7
×  1
```

38
```
   3
×  7
```

39
```
  12
×  7
```

40
```
   7
×  6
```

41
```
   7
×  8
```

42
```
   9
×  7
```

43
```
   1
×  7
```

44
```
   2
×  7
```

45
```
   6
×  7
```

46
```
   7
× 11
```

47
```
   8
×  7
```

48
```
   7
×  9
```

49
```
   7
×  0
```

50
```
   4
×  7
```

51
```
   8
×  7
```

52
```
   7
×  3
```

53
```
   7
× 11
```

54
```
   7
×  1
```

55
```
   1
×  7
```

56
```
   7
×  6
```

57
```
   9
×  7
```

58
```
   7
×  0
```

59
```
   7
×  9
```

60
```
   5
×  7
```

© 2024 Teach Elementals, LLC

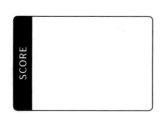

Name: _____   Date: _____

SCORE

1
7
× 9

2
7
×12

3
8
× 7

4
7
×11

5
7
× 8

6
11
× 7

7
2
× 7

8
7
× 2

9
4
× 7

10
7
× 6

11
7
× 1

12
0
× 7

13
9
× 7

14
7
× 0

15
7
× 3

16
7
×10

17
7
× 4

18
7
× 7

19
1
× 7

20
3
× 7

21
5
× 7

22
7
× 5

23
10
× 7

24
6
× 7

25
12
× 7

26
10
× 7

27
12
× 7

28
4
× 7

29
5
× 7

30
1
× 7

31
3
× 7

32
7
× 0

33
7
× 7

34
6
× 7

35
7
× 2

36
9
× 7

37
7
× 4

38
7
×10

39
7
×11

40
7
× 3

41
8
× 7

42
7
×12

43
7
× 8

44
7
× 6

45
7
× 5

46
0
× 7

47
7
× 1

48
2
× 7

49
11
× 7

50
7
× 9

51
7
× 8

52
7
× 0

53
7
× 7

54
11
× 7

55
5
× 7

56
1
× 7

57
7
× 4

58
7
× 6

59
7
×10

60
12
× 7

© 2024 Teach Elementals, LLC

Name: _____ Date: _____

SCORE

1
7
× 5

2
7
× 2

3
9
× 7

4
10
× 7

5
1
× 7

6
0
× 7

7
7
×10

8
11
× 7

9
2
× 7

10
4
× 7

11
7
×12

12
6
× 7

13
3
× 7

14
7
× 3

15
8
× 7

16
7
× 0

17
7
× 7

18
5
× 7

19
7
×11

20
7
× 1

21
7
× 8

22
7
× 9

23
12
× 7

24
7
× 6

25
7
× 4

26
7
× 3

27
7
× 1

28
4
× 7

29
7
× 0

30
7
×10

31
7
× 8

32
0
× 7

33
10
× 7

34
7
× 7

35
7
×11

36
7
× 9

37
9
× 7

38
7
×12

39
8
× 7

40
7
× 2

41
7
× 4

42
2
× 7

43
3
× 7

44
12
× 7

45
6
× 7

46
7
× 5

47
11
× 7

48
5
× 7

49
1
× 7

50
7
× 6

51
5
× 7

52
7
× 0

53
7
× 1

54
12
× 7

55
7
×12

56
7
× 2

57
0
× 7

58
11
× 7

59
7
×10

60
7
× 3

© 2024 Teach Elementals, LLC

Name: _____   Date: _____

SCORE

1 8 × 0	**2** 8 × 5	**3** 2 × 8	**4** 12 × 8

5 8 × 3	**6** 6 × 8	**7** 8 × 9	**8** 0 × 8	**9** 5 × 8	**10** 7 × 8	**11** 8 × 2	**12** 3 × 8
13 1 × 8	**14** 8 × 6	**15** 11 × 8	**16** 8 × 1	**17** 8 × 7	**18** 10 × 8	**19** 8 ×12	**20** 8 × 8
21 8 ×11	**22** 9 × 8	**23** 8 × 4	**24** 8 ×10	**25** 4 × 8	**26** 2 × 8	**27** 8 ×10	**28** 8 × 5
29 7 × 8	**30** 10 × 8	**31** 8 × 2	**32** 8 × 4	**33** 6 × 8	**34** 8 × 8	**35** 8 ×11	**36** 8 × 3
37 9 × 8	**38** 8 × 7	**39** 8 × 9	**40** 5 × 8	**41** 8 × 6	**42** 3 × 8	**43** 0 × 8	**44** 12 × 8
45 8 × 1	**46** 1 × 8	**47** 11 × 8	**48** 4 × 8	**49** 8 × 0	**50** 8 ×12	**51** 9 × 8	**52** 7 × 8
53 0 × 8	**54** 8 ×10	**55** 8 ×12	**56** 8 × 8	**57** 8 × 2	**58** 8 × 3	**59** 5 × 8	**60** 8 × 5

© 2024 Teach Elementals, LLC

Name: _____   **Date:** _____

SCORE

1	8 × 6
2	8 × 11
3	1 × 8
4	8 × 5

5	2 × 8
6	8 × 9
7	9 × 8
8	8 × 1
9	8 × 2
10	7 × 8
11	0 × 8
12	12 × 8

13	5 × 8
14	4 × 8
15	6 × 8
16	8 × 10
17	8 × 0
18	8 × 3
19	8 × 8
20	10 × 8

21	8 × 4
22	8 × 7
23	11 × 8
24	8 × 12
25	3 × 8
26	8 × 0
27	8 × 4
28	8 × 9

29	6 × 8
30	3 × 8
31	8 × 8
32	4 × 8
33	8 × 2
34	12 × 8
35	8 × 10
36	2 × 8

37	1 × 8
38	8 × 12
39	11 × 8
40	10 × 8
41	5 × 8
42	8 × 3
43	8 × 1
44	8 × 5

45	0 × 8
46	9 × 8
47	8 × 7
48	8 × 11
49	8 × 6
50	7 × 8
51	8 × 6
52	0 × 8

53	8 × 1
54	8 × 11
55	2 × 8
56	10 × 8
57	8 × 9
58	6 × 8
59	8 × 0
60	8 × 8

© 2024 Teach Elementals, LLC

SKILL SUPPORT

Name: _____

Date: _____

SCORE

1	**2**	**3**	**4**
8 × 7	3 × 8	8 × 11	8 × 9

5	**6**	**7**	**8**	**9**	**10**	**11**	**12**
8 × 4	8 × 12	8 × 5	12 × 8	8 × 10	5 × 8	8 × 6	1 × 8
13	**14**	**15**	**16**	**17**	**18**	**19**	**20**
2 × 8	9 × 8	4 × 8	0 × 8	7 × 8	8 × 8	6 × 8	8 × 0
21	**22**	**23**	**24**	**25**	**26**	**27**	**28**
8 × 1	10 × 8	11 × 8	8 × 3	8 × 2	8 × 4	6 × 8	8 × 6
29	**30**	**31**	**32**	**33**	**34**	**35**	**36**
8 × 2	8 × 10	8 × 0	7 × 8	8 × 5	8 × 3	12 × 8	8 × 11
37	**38**	**39**	**40**	**41**	**42**	**43**	**44**
8 × 9	0 × 8	1 × 8	8 × 1	9 × 8	8 × 8	8 × 12	10 × 8
45	**46**	**47**	**48**	**49**	**50**	**51**	**52**
5 × 8	4 × 8	8 × 7	3 × 8	11 × 8	2 × 8	11 × 8	8 × 7
53	**54**	**55**	**56**	**57**	**58**	**59**	**60**
6 × 8	8 × 2	8 × 0	4 × 8	0 × 8	8 × 12	9 × 8	8 × 10

© 2024 Teach Elementals, LLC

Name: _____ Date: _____

SCORE

1 8 × 0	**2** 1 × 8	**3** 8 × 5	**4** 8 × 6

5 8 ×12	**6** 8 × 4	**7** 10 × 8	**8** 8 ×11	**9** 9 × 8	**10** 2 × 8	**11** 8 × 3	**12** 7 × 8
13 8 × 1	**14** 8 × 8	**15** 8 × 9	**16** 8 × 2	**17** 0 × 8	**18** 8 × 7	**19** 6 × 8	**20** 11 × 8
21 3 × 8	**22** 5 × 8	**23** 8 ×10	**24** 12 × 8	**25** 4 × 8	**26** 8 ×12	**27** 8 × 5	**28** 8 ×10
29 4 × 8	**30** 7 × 8	**31** 8 × 4	**32** 8 × 1	**33** 11 × 8	**34** 8 × 0	**35** 8 × 2	**36** 8 × 6
37 9 × 8	**38** 0 × 8	**39** 2 × 8	**40** 10 × 8	**41** 6 × 8	**42** 8 × 9	**43** 5 × 8	**44** 12 × 8
45 8 × 3	**46** 8 × 8	**47** 1 × 8	**48** 8 ×11	**49** 8 × 7	**50** 3 × 8	**51** 8 × 8	**52** 8 × 1
53 8 ×11	**54** 4 × 8	**55** 8 × 7	**56** 8 × 6	**57** 8 × 5	**58** 12 × 8	**59** 0 × 8	**60** 8 × 2

© 2024 Teach Elementals, LLC

SKILL SUPPORT

Name: _____

Date: _____

SCORE

1 10 × 9	**2** 9 × 8	**3** 9 × 1	**4** 5 × 9

5 12 × 9	**6** 9 ×10	**7** 7 × 9	**8** 9 × 2	**9** 2 × 9	**10** 9 ×12	**11** 11 × 9	**12** 3 × 9
13 9 × 6	**14** 0 × 9	**15** 4 × 9	**16** 9 × 3	**17** 9 × 5	**18** 9 ×11	**19** 9 × 7	**20** 9 × 0
21 6 × 9	**22** 1 × 9	**23** 9 × 4	**24** 9 × 9	**25** 8 × 9	**26** 5 × 9	**27** 11 × 9	**28** 4 × 9
29 9 × 2	**30** 9 × 4	**31** 1 × 9	**32** 9 × 9	**33** 9 × 3	**34** 9 × 6	**35** 8 × 9	**36** 9 ×10
37 10 × 9	**38** 7 × 9	**39** 9 ×11	**40** 9 × 7	**41** 9 × 0	**42** 9 × 1	**43** 0 × 9	**44** 3 × 9
45 2 × 9	**46** 9 × 8	**47** 9 ×12	**48** 9 × 5	**49** 6 × 9	**50** 12 × 9	**51** 2 × 9	**52** 9 × 6
53 0 × 9	**54** 9 × 4	**55** 7 × 9	**56** 6 × 9	**57** 9 × 7	**58** 9 × 5	**59** 3 × 9	**60** 12 × 9

© 2024 Teach Elementals, LLC

Name: _____   Date: _____

SCORE

1
9
× 0

2
9
× 6

3
0
× 9

4
9
×12

5
9
× 9

6
5
× 9

7
10
× 9

8
9
× 7

9
12
× 9

10
9
× 5

11
9
× 4

12
1
× 9

13
4
× 9

14
11
× 9

15
9
× 3

16
9
× 1

17
9
×11

18
8
× 9

19
2
× 9

20
3
× 9

21
9
×10

22
7
× 9

23
6
× 9

24
9
× 8

25
9
× 2

26
6
× 9

27
3
× 9

28
9
× 9

29
10
× 9

30
9
× 4

31
9
× 0

32
9
× 2

33
9
× 3

34
9
×11

35
11
× 9

36
5
× 9

37
9
× 8

38
8
× 9

39
9
×10

40
9
×12

41
9
× 1

42
7
× 9

43
9
× 7

44
9
× 6

45
0
× 9

46
2
× 9

47
1
× 9

48
12
× 9

49
4
× 9

50
9
× 5

51
9
× 7

52
9
× 8

53
5
× 9

54
9
× 1

55
9
× 2

56
1
× 9

57
3
× 9

58
9
× 5

59
9
×10

60
0
× 9

© 2024 Teach Elementals, LLC

Name: _____   Date: _____

SCORE

1
  8
× 9

2
  9
× 0

3
  9
× 5

4
  9
× 2

5
  9
× 3

6
 10
× 9

7
  9
× 1

8
  9
× 8

9
  3
× 9

10
  9
×11

11
  9
×12

12
  4
× 9

13
  9
×10

14
  9
× 6

15
  6
× 9

16
  9
× 9

17
  0
× 9

18
  5
× 9

19
 11
× 9

20
  9
× 7

21
 12
× 9

22
  7
× 9

23
  2
× 9

24
  1
× 9

25
  9
× 4

26
  7
× 9

27
  6
× 9

28
  9
×10

29
  9
× 1

30
  4
× 9

31
  9
×11

32
  0
× 9

33
  9
× 2

34
 12
× 9

35
  9
× 4

36
  5
× 9

37
  9
× 0

38
  9
×12

39
  2
× 9

40
  9
× 5

41
  3
× 9

42
  8
× 9

43
  9
× 3

44
 10
× 9

45
  9
× 9

46
  9
× 6

47
  9
× 7

48
  1
× 9

49
 11
× 9

50
  9
× 8

51
 12
× 9

52
  1
× 9

53
  3
× 9

54
 10
× 9

55
  9
×11

56
  9
× 5

57
  2
× 9

58
  4
× 9

59
  9
×12

60
  9
× 8

© 2024 Teach Elementals, LLC

Name: _____   Date: _____

SCORE

1 4 × 9	**2** 11 × 9	**3** 3 × 9	**4** 9 × 8

5 8 × 9	**6** 2 × 9	**7** 9 × 7	**8** 9 × 3	**9** 0 × 9	**10** 9 × 6	**11** 9 × 9	**12** 7 × 9
13 9 × 0	**14** 5 × 9	**15** 9 × 4	**16** 9 × 5	**17** 9 × 1	**18** 10 × 9	**19** 6 × 9	**20** 9 ×11
21 9 ×10	**22** 9 ×12	**23** 1 × 9	**24** 12 × 9	**25** 9 × 2	**26** 7 × 9	**27** 9 ×11	**28** 9 × 7
29 9 × 6	**30** 9 × 8	**31** 9 × 5	**32** 0 × 9	**33** 9 × 1	**34** 5 × 9	**35** 2 × 9	**36** 6 × 9
37 4 × 9	**38** 9 × 4	**39** 9 ×10	**40** 9 × 2	**41** 9 × 3	**42** 9 × 0	**43** 11 × 9	**44** 10 × 9
45 3 × 9	**46** 9 ×12	**47** 8 × 9	**48** 9 × 9	**49** 1 × 9	**50** 12 × 9	**51** 9 × 8	**52** 9 × 7
53 4 × 9	**54** 9 × 0	**55** 6 × 9	**56** 9 × 9	**57** 2 × 9	**58** 10 × 9	**59** 9 × 5	**60** 3 × 9

© 2024 Teach Elementals, LLC

SKILL SUPPORT

Name: _____ Date: _____

SCORE

1
5
×10

2
1
×10

3
10
× 1

4
2
×10

5
8
×10

6
9
×10

7
10
× 6

8
11
×10

9
0
×10

10
10
× 8

11
10
× 5

12
10
×11

13
10
× 7

14
3
×10

15
6
×10

16
12
×10

17
4
×10

18
10
×12

19
7
×10

20
10
× 4

21
10
× 3

22
10
× 9

23
10
× 2

24
10
×10

25
10
× 0

26
10
× 2

27
0
×10

28
3
×10

29
10
× 3

30
12
×10

31
10
× 5

32
10
× 4

33
10
×10

34
4
×10

35
10
× 0

36
9
×10

37
6
×10

38
2
×10

39
5
×10

40
1
×10

41
10
× 9

42
10
× 1

43
10
×12

44
10
× 8

45
8
×10

46
7
×10

47
11
×10

48
10
× 7

49
10
×11

50
10
× 6

51
12
×10

52
4
×10

53
10
× 0

54
10
×10

55
10
×11

56
1
×10

57
10
× 9

58
5
×10

59
10
× 1

60
8
×10

© 2024 Teach Elementals, LLC

Name: _____   Date: _____

SCORE

1
10
× 7

2
2
×10

3
10
× 5

4
6
×10

5
7
×10

6
10
× 0

7
10
× 2

8
10
× 3

9
10
×11

10
10
× 8

11
9
×10

12
10
× 1

13
1
×10

14
11
×10

15
0
×10

16
8
×10

17
10
×10

18
4
×10

19
10
×12

20
10
× 4

21
3
×10

22
10
× 9

23
12
×10

24
10
× 6

25
5
×10

26
9
×10

27
10
× 5

28
10
× 6

29
10
× 8

30
11
×10

31
6
×10

32
5
×10

33
4
×10

34
1
×10

35
10
× 4

36
2
×10

37
10
× 7

38
10
× 0

39
10
× 1

40
12
×10

41
3
×10

42
10
×10

43
10
× 2

44
10
×11

45
0
×10

46
7
×10

47
10
× 9

48
10
×12

49
10
× 3

50
8
×10

51
1
×10

52
10
× 8

53
10
× 0

54
10
× 1

55
0
×10

56
7
×10

57
10
× 4

58
10
× 3

59
10
× 6

60
10
× 2

© 2024 Teach Elementals, LLC

Name: _____ **Date:** _____

SCORE

1. 11 × 10
2. 10 × 1
3. 10 × 5
4. 10 × 11

5. 3 × 10
6. 4 × 10
7. 10 × 10
8. 10 × 6
9. 10 × 0
10. 2 × 10
11. 7 × 10
12. 10 × 8

13. 9 × 10
14. 10 × 2
15. 6 × 10
16. 1 × 10
17. 12 × 10
18. 10 × 3
19. 8 × 10
20. 0 × 10

21. 10 × 12
22. 10 × 9
23. 10 × 4
24. 10 × 7
25. 5 × 10
26. 10 × 11
27. 10 × 9
28. 9 × 10

29. 8 × 10
30. 10 × 8
31. 12 × 10
32. 10 × 2
33. 10 × 12
34. 10 × 0
35. 10 × 7
36. 0 × 10

37. 10 × 5
38. 10 × 10
39. 10 × 6
40. 11 × 10
41. 6 × 10
42. 4 × 10
43. 10 × 4
44. 3 × 10

45. 2 × 10
46. 7 × 10
47. 10 × 1
48. 10 × 3
49. 5 × 10
50. 1 × 10
51. 4 × 10
52. 5 × 10

53. 10 × 1
54. 10 × 0
55. 9 × 10
56. 10 × 7
57. 1 × 10
58. 10 × 5
59. 10 × 4
60. 10 × 10

© 2024 Teach Elementals, LLC

Name: _____

Date: _____

SCORE

1 11 ×10	**2** 10 × 9	**3** 10 × 2	**4** 10 × 1
5 2 ×10	**6** 10 × 7	**7** 10 × 8	**8** 10 × 4
9 1 ×10	**10** 12 ×10	**11** 10 × 3	**12** 10 ×10
13 10 × 0	**14** 0 ×10	**15** 3 ×10	**16** 4 ×10
17 10 × 6	**18** 5 ×10	**19** 10 × 5	**20** 10 ×11
21 7 ×10	**22** 10 ×12	**23** 9 ×10	**24** 6 ×10
25 8 ×10	**26** 10 × 9	**27** 10 × 3	**28** 10 × 2
29 8 ×10	**30** 2 ×10	**31** 10 × 5	**32** 11 ×10
33 6 ×10	**34** 5 ×10	**35** 10 ×10	**36** 10 × 1
37 12 ×10	**38** 3 ×10	**39** 10 ×11	**40** 1 ×10
41 7 ×10	**42** 0 ×10	**43** 10 ×12	**44** 10 × 0
45 10 × 6	**46** 10 × 4	**47** 10 × 8	**48** 4 ×10
49 9 ×10	**50** 10 × 7	**51** 10 × 2	**52** 1 ×10
53 4 ×10	**54** 9 ×10	**55** 10 × 4	**56** 5 ×10
57 2 ×10	**58** 10 ×10	**59** 11 ×10	**60** 10 × 5

SKILL SUPPORT

Name: _____

Date: _____

SCORE

1) 11 × 5	2) 10 ×11
3) 5 ×11	4) 11 × 1

5) 11 ×11

6) 11 × 0

7) 4 ×11

8) 7 ×11

9) 11 × 4

10) 1 ×11

11) 11 × 7

12) 11 × 9

13) 11 ×12

14) 9 ×11

15) 2 ×11

16) 11 × 2

17) 3 ×11

18) 11 × 6

19) 8 ×11

20) 0 ×11

21) 6 ×11

22) 11 × 8

23) 12 ×11

24) 11 ×10

25) 11 × 3

26) 5 ×11

27) 11 × 3

28) 11 × 2

29) 11 ×12

30) 0 ×11

31) 2 ×11

32) 10 ×11

33) 11 × 7

34) 11 × 4

35) 11 × 8

36) 8 ×11

37) 11 × 0

38) 3 ×11

39) 4 ×11

40) 11 × 5

41) 11 × 1

42) 11 ×10

43) 11 × 9

44) 9 ×11

45) 11 × 6

46) 1 ×11

47) 11 ×11

48) 7 ×11

49) 12 ×11

50) 6 ×11

51) 11 × 3

52) 2 ×11

53) 12 ×11

54) 4 ×11

55) 11 ×12

56) 11 × 0

57) 11 × 9

58) 7 ×11

59) 11 × 4

60) 6 ×11

© 2024 Teach Elementals, LLC

Name: _____ Date: _____

SCORE

1 $\begin{array}{r} 12 \\ \times 11 \\ \hline \end{array}$	**2** $\begin{array}{r} 8 \\ \times 11 \\ \hline \end{array}$	**3** $\begin{array}{r} 11 \\ \times 11 \\ \hline \end{array}$	**4** $\begin{array}{r} 4 \\ \times 11 \\ \hline \end{array}$

5 $\begin{array}{r} 11 \\ \times\ 3 \\ \hline \end{array}$	**6** $\begin{array}{r} 11 \\ \times\ 8 \\ \hline \end{array}$	**7** $\begin{array}{r} 9 \\ \times 11 \\ \hline \end{array}$	**8** $\begin{array}{r} 11 \\ \times\ 9 \\ \hline \end{array}$	**9** $\begin{array}{r} 3 \\ \times 11 \\ \hline \end{array}$	**10** $\begin{array}{r} 11 \\ \times 10 \\ \hline \end{array}$	**11** $\begin{array}{r} 10 \\ \times 11 \\ \hline \end{array}$	**12** $\begin{array}{r} 11 \\ \times 12 \\ \hline \end{array}$
13 $\begin{array}{r} 11 \\ \times\ 0 \\ \hline \end{array}$	**14** $\begin{array}{r} 11 \\ \times\ 2 \\ \hline \end{array}$	**15** $\begin{array}{r} 2 \\ \times 11 \\ \hline \end{array}$	**16** $\begin{array}{r} 11 \\ \times\ 4 \\ \hline \end{array}$	**17** $\begin{array}{r} 11 \\ \times\ 6 \\ \hline \end{array}$	**18** $\begin{array}{r} 7 \\ \times 11 \\ \hline \end{array}$	**19** $\begin{array}{r} 6 \\ \times 11 \\ \hline \end{array}$	**20** $\begin{array}{r} 0 \\ \times 11 \\ \hline \end{array}$
21 $\begin{array}{r} 5 \\ \times 11 \\ \hline \end{array}$	**22** $\begin{array}{r} 11 \\ \times\ 1 \\ \hline \end{array}$	**23** $\begin{array}{r} 11 \\ \times\ 7 \\ \hline \end{array}$	**24** $\begin{array}{r} 1 \\ \times 11 \\ \hline \end{array}$	**25** $\begin{array}{r} 11 \\ \times\ 5 \\ \hline \end{array}$	**26** $\begin{array}{r} 5 \\ \times 11 \\ \hline \end{array}$	**27** $\begin{array}{r} 11 \\ \times\ 1 \\ \hline \end{array}$	**28** $\begin{array}{r} 11 \\ \times\ 7 \\ \hline \end{array}$
29 $\begin{array}{r} 2 \\ \times 11 \\ \hline \end{array}$	**30** $\begin{array}{r} 10 \\ \times 11 \\ \hline \end{array}$	**31** $\begin{array}{r} 11 \\ \times 11 \\ \hline \end{array}$	**32** $\begin{array}{r} 11 \\ \times 12 \\ \hline \end{array}$	**33** $\begin{array}{r} 9 \\ \times 11 \\ \hline \end{array}$	**34** $\begin{array}{r} 7 \\ \times 11 \\ \hline \end{array}$	**35** $\begin{array}{r} 3 \\ \times 11 \\ \hline \end{array}$	**36** $\begin{array}{r} 11 \\ \times\ 5 \\ \hline \end{array}$
37 $\begin{array}{r} 6 \\ \times 11 \\ \hline \end{array}$	**38** $\begin{array}{r} 11 \\ \times\ 0 \\ \hline \end{array}$	**39** $\begin{array}{r} 1 \\ \times 11 \\ \hline \end{array}$	**40** $\begin{array}{r} 11 \\ \times\ 4 \\ \hline \end{array}$	**41** $\begin{array}{r} 11 \\ \times 10 \\ \hline \end{array}$	**42** $\begin{array}{r} 11 \\ \times\ 3 \\ \hline \end{array}$	**43** $\begin{array}{r} 11 \\ \times\ 2 \\ \hline \end{array}$	**44** $\begin{array}{r} 12 \\ \times 11 \\ \hline \end{array}$
45 $\begin{array}{r} 4 \\ \times 11 \\ \hline \end{array}$	**46** $\begin{array}{r} 8 \\ \times 11 \\ \hline \end{array}$	**47** $\begin{array}{r} 11 \\ \times\ 8 \\ \hline \end{array}$	**48** $\begin{array}{r} 0 \\ \times 11 \\ \hline \end{array}$	**49** $\begin{array}{r} 11 \\ \times\ 6 \\ \hline \end{array}$	**50** $\begin{array}{r} 11 \\ \times\ 9 \\ \hline \end{array}$	**51** $\begin{array}{r} 0 \\ \times 11 \\ \hline \end{array}$	**52** $\begin{array}{r} 7 \\ \times 11 \\ \hline \end{array}$
53 $\begin{array}{r} 8 \\ \times 11 \\ \hline \end{array}$	**54** $\begin{array}{r} 11 \\ \times 10 \\ \hline \end{array}$	**55** $\begin{array}{r} 5 \\ \times 11 \\ \hline \end{array}$	**56** $\begin{array}{r} 1 \\ \times 11 \\ \hline \end{array}$	**57** $\begin{array}{r} 4 \\ \times 11 \\ \hline \end{array}$	**58** $\begin{array}{r} 11 \\ \times\ 1 \\ \hline \end{array}$	**59** $\begin{array}{r} 12 \\ \times 11 \\ \hline \end{array}$	**60** $\begin{array}{r} 11 \\ \times\ 8 \\ \hline \end{array}$

© 2024 Teach Elementals, LLC

Name: _____

Date: _____

SCORE

1
  8
×11

2
 11
× 7

3
  4
×11

4
 11
× 3

5
 12
×11

6
 11
× 0

7
  6
×11

8
  7
×11

9
 11
×12

10
  5
×11

11
  0
×11

12
 11
× 1

13
 11
× 4

14
 10
×11

15
  9
×11

16
 11
×10

17
 11
× 6

18
  3
×11

19
 11
×11

20
 11
× 5

21
 11
× 8

22
 11
× 9

23
  1
×11

24
  2
×11

25
 11
× 2

26
 11
× 7

27
  2
×11

28
 11
×12

29
 11
×10

30
  5
×11

31
 11
× 2

32
  3
×11

33
 11
× 4

34
 11
× 3

35
 11
× 5

36
  7
×11

37
 11
× 0

38
 10
×11

39
 12
×11

40
 11
× 1

41
  0
×11

42
 11
× 6

43
  4
×11

44
  8
×11

45
  9
×11

46
  1
×11

47
 11
× 8

48
 11
×11

49
 11
× 9

50
  6
×11

51
 11
× 3

52
  7
×11

53
  9
×11

54
 11
× 4

55
 11
×11

56
 11
× 6

57
 12
×11

58
  3
×11

59
  8
×11

60
 11
× 9

© 2024 Teach Elementals, LLC

Name: _____   Date: _____

SCORE

1
11
× 1

2
8
×11

3
9
×11

4
10
×11

5
11
× 0

6
7
×11

7
1
×11

8
5
×11

9
2
×11

10
11
×10

11
11
× 8

12
11
× 5

13
11
×12

14
11
× 3

15
6
×11

16
3
×11

17
11
× 7

18
0
×11

19
11
× 4

20
11
× 6

21
11
× 9

22
11
×11

23
11
× 2

24
12
×11

25
4
×11

26
3
×11

27
11
× 7

28
11
× 5

29
11
× 1

30
11
× 3

31
4
×11

32
2
×11

33
12
×11

34
10
×11

35
11
× 2

36
0
×11

37
11
× 8

38
1
×11

39
9
×11

40
6
×11

41
11
×10

42
11
× 9

43
11
× 4

44
7
×11

45
5
×11

46
8
×11

47
11
× 0

48
11
× 6

49
11
×12

50
11
×11

51
11
× 1

52
9
×11

53
6
×11

54
10
×11

55
0
×11

56
2
×11

57
11
× 5

58
11
× 2

59
11
×10

60
5
×11

© 2024 Teach Elementals, LLC

MULTIPLYING BY 12, SET A

Name: _____   Date: _____

SCORE

1. 11
    ×12

2. 12
    × 9

3.   4
    ×12

4.   9
    ×12

5. 12
    × 8

6.   8
    ×12

7.   2
    ×12

8. 12
    × 1

9. 12
    ×10

10. 12
    × 5

11. 3
    ×12

12. 12
    × 6

13. 12
    × 4

14. 12
    ×12

15. 12
    × 0

16. 12
    × 3

17. 10
    ×12

18. 1
    ×12

19. 12
    ×11

20. 7
    ×12

21. 5
    ×12

22. 0
    ×12

23. 12
    × 7

24. 12
    × 2

25. 6
    ×12

26. 0
    ×12

27. 7
    ×12

28. 12
    × 4

29. 8
    ×12

30. 2
    ×12

31. 12
    × 2

32. 12
    × 7

33. 12
    × 1

34. 11
    ×12

35. 12
    ×12

36. 1
    ×12

37. 12
    ×11

38. 12
    × 3

39. 12
    ×10

40. 3
    ×12

41. 6
    ×12

42. 12
    × 8

43. 10
    ×12

44. 12
    × 0

45. 5
    ×12

46. 12
    × 9

47. 9
    ×12

48. 12
    × 5

49. 12
    × 6

50. 4
    ×12

51. 10
    ×12

52. 12
    × 9

53. 1
    ×12

54. 3
    ×12

55. 12
    × 3

56. 0
    ×12

57. 11
    ×12

58. 2
    ×12

59. 12
    ×10

60. 12
    × 7

© 2024 Teach Elementals, LLC

Name: _____   Date: _____

SCORE

1 12 × 1	**2** 12 × 6	**3** 12 ×11	**4** 12 × 7
5 9 ×12	**6** 11 ×12	**7** 12 × 8	**8** 2 ×12
9 12 × 3	**10** 12 × 4	**11** 12 × 0	**12** 12 × 2
13 3 ×12	**14** 5 ×12	**15** 12 ×10	**16** 6 ×12
17 8 ×12	**18** 0 ×12	**19** 4 ×12	**20** 12 × 5
21 12 ×12	**22** 1 ×12	**23** 7 ×12	**24** 10 ×12
25 12 × 9	**26** 12 ×10	**27** 12 × 1	**28** 1 ×12
29 12 ×11	**30** 12 × 9	**31** 5 ×12	**32** 11 ×12
33 12 × 7	**34** 12 × 2	**35** 3 ×12	**36** 9 ×12
37 4 ×12	**38** 12 × 3	**39** 12 ×12	**40** 2 ×12
41 8 ×12	**42** 12 × 6	**43** 10 ×12	**44** 12 × 5
45 12 × 8	**46** 0 ×12	**47** 7 ×12	**48** 12 × 4
49 12 × 0	**50** 6 ×12	**51** 12 × 4	**52** 12 × 3
53 6 ×12	**54** 12 × 7	**55** 7 ×12	**56** 3 ×12
57 12 × 2	**58** 10 ×12	**59** 8 ×12	**60** 12 ×12

© 2024 Teach Elementals, LLC

Name: _____   Date: _____

SCORE

1
  2
×12

2
  7
×12

3
 12
×10

4
  9
×12

5
 12
× 2

6
 12
× 1

7
  0
×12

8
  6
×12

9
 12
× 9

10
  4
×12

11
  8
×12

12
  1
×12

13
  3
×12

14
  5
×12

15
 12
× 5

16
 12
× 6

17
 12
× 3

18
 12
× 7

19
 10
×12

20
 12
× 0

21
 12
×12

22
 12
× 4

23
 12
×11

24
 12
× 8

25
 11
×12

26
 12
× 4

27
  5
×12

28
 10
×12

29
 12
× 5

30
 12
× 6

31
  3
×12

32
 11
×12

33
 12
× 1

34
  6
×12

35
 12
× 8

36
 12
× 9

37
 12
× 7

38
  1
×12

39
 12
×10

40
  7
×12

41
 12
×12

42
 12
× 3

43
  4
×12

44
 12
× 0

45
 12
× 2

46
  0
×12

47
  9
×12

48
  8
×12

49
 12
×11

50
  2
×12

51
  1
×12

52
  2
×12

53
 12
× 1

54
 12
× 2

55
 12
×11

56
  4
×12

57
 12
× 6

58
  9
×12

59
 12
× 8

60
 11
×12

© 2024 Teach Elementals, LLC

Name: _____

Date: _____

SCORE

1 12 × 7	**2** 9 ×12	**3** 12 × 8	**4** 11 ×12
5 6 ×12	**6** 12 × 2	**7** 12 × 9	**8** 5 ×12
9 2 ×12	**10** 4 ×12	**11** 1 ×12	**12** 12 × 4
13 0 ×12	**14** 12 × 6	**15** 12 ×10	**16** 12 × 1
17 12 × 5	**18** 12 ×12	**19** 8 ×12	**20** 10 ×12
21 7 ×12	**22** 3 ×12	**23** 12 × 3	**24** 12 ×11
25 12 × 0	**26** 12 × 4	**27** 0 ×12	**28** 9 ×12
29 12 × 7	**30** 12 ×12	**31** 12 × 8	**32** 3 ×12
33 11 ×12	**34** 4 ×12	**35** 12 ×10	**36** 5 ×12
37 12 ×11	**38** 12 × 9	**39** 12 × 1	**40** 7 ×12
41 1 ×12	**42** 12 × 5	**43** 10 ×12	**44** 2 ×12
45 12 × 0	**46** 8 ×12	**47** 12 × 2	**48** 6 ×12
49 12 × 6	**50** 12 × 3	**51** 12 × 5	**52** 11 ×12
53 12 × 3	**54** 1 ×12	**55** 12 ×11	**56** 2 ×12
57 12 ×10	**58** 7 ×12	**59** 12 × 1	**60** 12 × 8

© 2024 Teach Elementals, LLC

MIXED REVIEW: 0 THROUGH 5, SET A

Name: _____   Date: _____

SCORE

1 $\begin{array}{r} 1 \\ \times 11 \\ \hline \end{array}$	**2** $\begin{array}{r} 1 \\ \times 12 \\ \hline \end{array}$	**3** $\begin{array}{r} 10 \\ \times 1 \\ \hline \end{array}$	**4** $\begin{array}{r} 6 \\ \times 4 \\ \hline \end{array}$

5 $\begin{array}{r} 5 \\ \times 1 \\ \hline \end{array}$	**6** $\begin{array}{r} 3 \\ \times 9 \\ \hline \end{array}$	**7** $\begin{array}{r} 2 \\ \times 11 \\ \hline \end{array}$	**8** $\begin{array}{r} 3 \\ \times 7 \\ \hline \end{array}$	**9** $\begin{array}{r} 2 \\ \times 6 \\ \hline \end{array}$	**10** $\begin{array}{r} 3 \\ \times 1 \\ \hline \end{array}$	**11** $\begin{array}{r} 5 \\ \times 3 \\ \hline \end{array}$	**12** $\begin{array}{r} 1 \\ \times 2 \\ \hline \end{array}$
13 $\begin{array}{r} 4 \\ \times 6 \\ \hline \end{array}$	**14** $\begin{array}{r} 5 \\ \times 10 \\ \hline \end{array}$	**15** $\begin{array}{r} 1 \\ \times 8 \\ \hline \end{array}$	**16** $\begin{array}{r} 3 \\ \times 4 \\ \hline \end{array}$	**17** $\begin{array}{r} 4 \\ \times 0 \\ \hline \end{array}$	**18** $\begin{array}{r} 5 \\ \times 4 \\ \hline \end{array}$	**19** $\begin{array}{r} 11 \\ \times 5 \\ \hline \end{array}$	**20** $\begin{array}{r} 7 \\ \times 3 \\ \hline \end{array}$
21 $\begin{array}{r} 5 \\ \times 0 \\ \hline \end{array}$	**22** $\begin{array}{r} 4 \\ \times 5 \\ \hline \end{array}$	**23** $\begin{array}{r} 4 \\ \times 3 \\ \hline \end{array}$	**24** $\begin{array}{r} 5 \\ \times 7 \\ \hline \end{array}$	**25** $\begin{array}{r} 4 \\ \times 5 \\ \hline \end{array}$	**26** $\begin{array}{r} 5 \\ \times 6 \\ \hline \end{array}$	**27** $\begin{array}{r} 8 \\ \times 4 \\ \hline \end{array}$	**28** $\begin{array}{r} 12 \\ \times 3 \\ \hline \end{array}$
29 $\begin{array}{r} 8 \\ \times 1 \\ \hline \end{array}$	**30** $\begin{array}{r} 2 \\ \times 5 \\ \hline \end{array}$	**31** $\begin{array}{r} 0 \\ \times 12 \\ \hline \end{array}$	**32** $\begin{array}{r} 0 \\ \times 1 \\ \hline \end{array}$	**33** $\begin{array}{r} 10 \\ \times 3 \\ \hline \end{array}$	**34** $\begin{array}{r} 8 \\ \times 0 \\ \hline \end{array}$	**35** $\begin{array}{r} 0 \\ \times 3 \\ \hline \end{array}$	**36** $\begin{array}{r} 1 \\ \times 1 \\ \hline \end{array}$
37 $\begin{array}{r} 3 \\ \times 11 \\ \hline \end{array}$	**38** $\begin{array}{r} 0 \\ \times 0 \\ \hline \end{array}$	**39** $\begin{array}{r} 6 \\ \times 0 \\ \hline \end{array}$	**40** $\begin{array}{r} 3 \\ \times 5 \\ \hline \end{array}$	**41** $\begin{array}{r} 7 \\ \times 2 \\ \hline \end{array}$	**42** $\begin{array}{r} 3 \\ \times 0 \\ \hline \end{array}$	**43** $\begin{array}{r} 12 \\ \times 4 \\ \hline \end{array}$	**44** $\begin{array}{r} 1 \\ \times 10 \\ \hline \end{array}$
45 $\begin{array}{r} 2 \\ \times 2 \\ \hline \end{array}$	**46** $\begin{array}{r} 11 \\ \times 1 \\ \hline \end{array}$	**47** $\begin{array}{r} 0 \\ \times 10 \\ \hline \end{array}$	**48** $\begin{array}{r} 3 \\ \times 4 \\ \hline \end{array}$	**49** $\begin{array}{r} 0 \\ \times 2 \\ \hline \end{array}$	**50** $\begin{array}{r} 2 \\ \times 0 \\ \hline \end{array}$	**51** $\begin{array}{r} 9 \\ \times 4 \\ \hline \end{array}$	**52** $\begin{array}{r} 4 \\ \times 12 \\ \hline \end{array}$
53 $\begin{array}{r} 3 \\ \times 1 \\ \hline \end{array}$	**54** $\begin{array}{r} 6 \\ \times 2 \\ \hline \end{array}$	**55** $\begin{array}{r} 1 \\ \times 5 \\ \hline \end{array}$	**56** $\begin{array}{r} 0 \\ \times 7 \\ \hline \end{array}$	**57** $\begin{array}{r} 5 \\ \times 8 \\ \hline \end{array}$	**58** $\begin{array}{r} 0 \\ \times 4 \\ \hline \end{array}$	**59** $\begin{array}{r} 0 \\ \times 8 \\ \hline \end{array}$	**60** $\begin{array}{r} 10 \\ \times 0 \\ \hline \end{array}$

© 2024 Teach Elementals, LLC

Name: _____ Date: _____

SCORE

1
3
× 2

2
6
× 3

3
0
× 6

4
0
× 1

5
1
× 6

6
2
× 9

7
9
× 3

8
1
× 4

9
12
× 0

10
11
× 4

11
6
× 5

12
4
× 2

13
9
× 1

14
8
× 2

15
8
× 3

16
1
× 4

17
2
× 1

18
7
× 4

19
8
× 5

20
0
× 2

21
1
× 9

22
5
× 2

23
5
× 1

24
0
× 3

25
4
×10

26
0
× 5

27
3
× 2

28
5
× 2

29
7
× 1

30
1
× 0

31
5
× 5

32
4
×11

33
2
× 8

34
4
× 1

35
7
× 5

36
5
×12

37
4
× 8

38
5
× 3

39
3
× 5

40
5
× 4

41
2
×12

42
4
× 9

43
4
× 4

44
10
× 2

45
3
×10

46
2
× 3

47
2
× 0

48
5
×11

49
0
× 5

50
9
× 5

51
12
× 1

52
2
× 4

53
2
× 7

54
1
× 2

55
3
× 8

56
2
×10

57
4
× 1

58
1
× 3

59
1
× 5

60
12
× 5

© 2024 Teach Elementals, LLC

MIXED REVIEW: 0 THROUGH 5, SET C

Name: _____ Date: _____

SCORE

1
$$\begin{array}{r} 10 \\ \times\ 5 \\ \hline \end{array}$$

2
$$\begin{array}{r} 2 \\ \times\ 2 \\ \hline \end{array}$$

3
$$\begin{array}{r} 4 \\ \times\ 3 \\ \hline \end{array}$$

4
$$\begin{array}{r} 1 \\ \times 10 \\ \hline \end{array}$$

5
$$\begin{array}{r} 11 \\ \times\ 0 \\ \hline \end{array}$$

6
$$\begin{array}{r} 2 \\ \times\ 1 \\ \hline \end{array}$$

7
$$\begin{array}{r} 5 \\ \times\ 9 \\ \hline \end{array}$$

8
$$\begin{array}{r} 4 \\ \times\ 7 \\ \hline \end{array}$$

9
$$\begin{array}{r} 7 \\ \times\ 2 \\ \hline \end{array}$$

10
$$\begin{array}{r} 1 \\ \times\ 2 \\ \hline \end{array}$$

11
$$\begin{array}{r} 4 \\ \times\ 0 \\ \hline \end{array}$$

12
$$\begin{array}{r} 6 \\ \times\ 0 \\ \hline \end{array}$$

13
$$\begin{array}{r} 4 \\ \times\ 2 \\ \hline \end{array}$$

14
$$\begin{array}{r} 1 \\ \times\ 7 \\ \hline \end{array}$$

15
$$\begin{array}{r} 12 \\ \times\ 5 \\ \hline \end{array}$$

16
$$\begin{array}{r} 2 \\ \times\ 7 \\ \hline \end{array}$$

17
$$\begin{array}{r} 3 \\ \times 11 \\ \hline \end{array}$$

18
$$\begin{array}{r} 10 \\ \times\ 3 \\ \hline \end{array}$$

19
$$\begin{array}{r} 2 \\ \times\ 0 \\ \hline \end{array}$$

20
$$\begin{array}{r} 9 \\ \times\ 5 \\ \hline \end{array}$$

21
$$\begin{array}{r} 9 \\ \times\ 2 \\ \hline \end{array}$$

22
$$\begin{array}{r} 2 \\ \times\ 4 \\ \hline \end{array}$$

23
$$\begin{array}{r} 0 \\ \times\ 5 \\ \hline \end{array}$$

24
$$\begin{array}{r} 1 \\ \times\ 0 \\ \hline \end{array}$$

25
$$\begin{array}{r} 8 \\ \times\ 0 \\ \hline \end{array}$$

26
$$\begin{array}{r} 2 \\ \times\ 3 \\ \hline \end{array}$$

27
$$\begin{array}{r} 2 \\ \times\ 3 \\ \hline \end{array}$$

28
$$\begin{array}{r} 3 \\ \times\ 0 \\ \hline \end{array}$$

29
$$\begin{array}{r} 12 \\ \times\ 2 \\ \hline \end{array}$$

30
$$\begin{array}{r} 0 \\ \times\ 3 \\ \hline \end{array}$$

31
$$\begin{array}{r} 11 \\ \times\ 2 \\ \hline \end{array}$$

32
$$\begin{array}{r} 3 \\ \times\ 6 \\ \hline \end{array}$$

33
$$\begin{array}{r} 3 \\ \times\ 5 \\ \hline \end{array}$$

34
$$\begin{array}{r} 5 \\ \times\ 0 \\ \hline \end{array}$$

35
$$\begin{array}{r} 0 \\ \times 12 \\ \hline \end{array}$$

36
$$\begin{array}{r} 0 \\ \times 11 \\ \hline \end{array}$$

37
$$\begin{array}{r} 0 \\ \times\ 0 \\ \hline \end{array}$$

38
$$\begin{array}{r} 3 \\ \times\ 3 \\ \hline \end{array}$$

39
$$\begin{array}{r} 3 \\ \times\ 0 \\ \hline \end{array}$$

40
$$\begin{array}{r} 1 \\ \times\ 5 \\ \hline \end{array}$$

41
$$\begin{array}{r} 2 \\ \times 10 \\ \hline \end{array}$$

42
$$\begin{array}{r} 12 \\ \times\ 4 \\ \hline \end{array}$$

43
$$\begin{array}{r} 3 \\ \times 12 \\ \hline \end{array}$$

44
$$\begin{array}{r} 6 \\ \times\ 1 \\ \hline \end{array}$$

45
$$\begin{array}{r} 10 \\ \times\ 4 \\ \hline \end{array}$$

46
$$\begin{array}{r} 2 \\ \times\ 5 \\ \hline \end{array}$$

47
$$\begin{array}{r} 1 \\ \times\ 1 \\ \hline \end{array}$$

48
$$\begin{array}{r} 12 \\ \times\ 1 \\ \hline \end{array}$$

49
$$\begin{array}{r} 11 \\ \times\ 3 \\ \hline \end{array}$$

50
$$\begin{array}{r} 0 \\ \times\ 4 \\ \hline \end{array}$$

51
$$\begin{array}{r} 4 \\ \times\ 1 \\ \hline \end{array}$$

52
$$\begin{array}{r} 1 \\ \times\ 3 \\ \hline \end{array}$$

53
$$\begin{array}{r} 1 \\ \times\ 3 \\ \hline \end{array}$$

54
$$\begin{array}{r} 0 \\ \times\ 9 \\ \hline \end{array}$$

55
$$\begin{array}{r} 7 \\ \times\ 0 \\ \hline \end{array}$$

56
$$\begin{array}{r} 9 \\ \times\ 0 \\ \hline \end{array}$$

57
$$\begin{array}{r} 5 \\ \times 11 \\ \hline \end{array}$$

58
$$\begin{array}{r} 0 \\ \times\ 1 \\ \hline \end{array}$$

59
$$\begin{array}{r} 2 \\ \times\ 4 \\ \hline \end{array}$$

60
$$\begin{array}{r} 3 \\ \times\ 8 \\ \hline \end{array}$$

© 2024 Teach Elementals, LLC

Name: _____

Date: _____

SCORE

1
$$\begin{array}{r} 10 \\ \times\ 8 \\ \hline \end{array}$$

2
$$\begin{array}{r} 9 \\ \times 12 \\ \hline \end{array}$$

3
$$\begin{array}{r} 1 \\ \times\ 6 \\ \hline \end{array}$$

4
$$\begin{array}{r} 0 \\ \times\ 7 \\ \hline \end{array}$$

5
$$\begin{array}{r} 11 \\ \times\ 2 \\ \hline \end{array}$$

6
$$\begin{array}{r} 1 \\ \times\ 9 \\ \hline \end{array}$$

7
$$\begin{array}{r} 6 \\ \times\ 9 \\ \hline \end{array}$$

8
$$\begin{array}{r} 11 \\ \times\ 9 \\ \hline \end{array}$$

9
$$\begin{array}{r} 7 \\ \times 12 \\ \hline \end{array}$$

10
$$\begin{array}{r} 9 \\ \times\ 4 \\ \hline \end{array}$$

11
$$\begin{array}{r} 3 \\ \times\ 6 \\ \hline \end{array}$$

12
$$\begin{array}{r} 12 \\ \times\ 1 \\ \hline \end{array}$$

13
$$\begin{array}{r} 6 \\ \times\ 8 \\ \hline \end{array}$$

14
$$\begin{array}{r} 5 \\ \times\ 6 \\ \hline \end{array}$$

15
$$\begin{array}{r} 10 \\ \times 11 \\ \hline \end{array}$$

16
$$\begin{array}{r} 3 \\ \times\ 7 \\ \hline \end{array}$$

17
$$\begin{array}{r} 7 \\ \times 11 \\ \hline \end{array}$$

18
$$\begin{array}{r} 8 \\ \times\ 0 \\ \hline \end{array}$$

19
$$\begin{array}{r} 12 \\ \times 11 \\ \hline \end{array}$$

20
$$\begin{array}{r} 9 \\ \times\ 7 \\ \hline \end{array}$$

21
$$\begin{array}{r} 12 \\ \times\ 8 \\ \hline \end{array}$$

22
$$\begin{array}{r} 2 \\ \times\ 8 \\ \hline \end{array}$$

23
$$\begin{array}{r} 11 \\ \times\ 6 \\ \hline \end{array}$$

24
$$\begin{array}{r} 7 \\ \times 10 \\ \hline \end{array}$$

25
$$\begin{array}{r} 8 \\ \times 12 \\ \hline \end{array}$$

26
$$\begin{array}{r} 7 \\ \times\ 4 \\ \hline \end{array}$$

27
$$\begin{array}{r} 5 \\ \times\ 9 \\ \hline \end{array}$$

28
$$\begin{array}{r} 10 \\ \times\ 9 \\ \hline \end{array}$$

29
$$\begin{array}{r} 10 \\ \times\ 0 \\ \hline \end{array}$$

30
$$\begin{array}{r} 8 \\ \times\ 4 \\ \hline \end{array}$$

31
$$\begin{array}{r} 8 \\ \times 11 \\ \hline \end{array}$$

32
$$\begin{array}{r} 11 \\ \times\ 4 \\ \hline \end{array}$$

33
$$\begin{array}{r} 6 \\ \times\ 8 \\ \hline \end{array}$$

34
$$\begin{array}{r} 8 \\ \times\ 6 \\ \hline \end{array}$$

35
$$\begin{array}{r} 9 \\ \times\ 9 \\ \hline \end{array}$$

36
$$\begin{array}{r} 6 \\ \times\ 5 \\ \hline \end{array}$$

37
$$\begin{array}{r} 9 \\ \times\ 6 \\ \hline \end{array}$$

38
$$\begin{array}{r} 9 \\ \times\ 8 \\ \hline \end{array}$$

39
$$\begin{array}{r} 12 \\ \times\ 6 \\ \hline \end{array}$$

40
$$\begin{array}{r} 0 \\ \times 10 \\ \hline \end{array}$$

41
$$\begin{array}{r} 1 \\ \times 12 \\ \hline \end{array}$$

42
$$\begin{array}{r} 10 \\ \times 11 \\ \hline \end{array}$$

43
$$\begin{array}{r} 7 \\ \times\ 8 \\ \hline \end{array}$$

44
$$\begin{array}{r} 6 \\ \times 11 \\ \hline \end{array}$$

45
$$\begin{array}{r} 8 \\ \times\ 3 \\ \hline \end{array}$$

46
$$\begin{array}{r} 10 \\ \times\ 6 \\ \hline \end{array}$$

47
$$\begin{array}{r} 8 \\ \times 11 \\ \hline \end{array}$$

48
$$\begin{array}{r} 4 \\ \times\ 9 \\ \hline \end{array}$$

49
$$\begin{array}{r} 5 \\ \times 10 \\ \hline \end{array}$$

50
$$\begin{array}{r} 12 \\ \times\ 5 \\ \hline \end{array}$$

51
$$\begin{array}{r} 8 \\ \times 10 \\ \hline \end{array}$$

52
$$\begin{array}{r} 2 \\ \times 10 \\ \hline \end{array}$$

53
$$\begin{array}{r} 12 \\ \times\ 2 \\ \hline \end{array}$$

54
$$\begin{array}{r} 6 \\ \times\ 4 \\ \hline \end{array}$$

55
$$\begin{array}{r} 9 \\ \times\ 7 \\ \hline \end{array}$$

56
$$\begin{array}{r} 0 \\ \times 11 \\ \hline \end{array}$$

57
$$\begin{array}{r} 9 \\ \times\ 2 \\ \hline \end{array}$$

58
$$\begin{array}{r} 9 \\ \times\ 3 \\ \hline \end{array}$$

59
$$\begin{array}{r} 12 \\ \times 10 \\ \hline \end{array}$$

60
$$\begin{array}{r} 8 \\ \times\ 8 \\ \hline \end{array}$$

© 2024 Teach Elementals, LLC

Name: _____

Date: _____

SCORE

1 11 × 5	**2** 6 × 6	**3** 2 ×12	**4** 11 × 0

5 10 × 3	**6** 9 ×10	**7** 7 × 1	**8** 12 ×10	**9** 11 ×12	**10** 3 × 8	**11** 11 × 6	**12** 9 × 8
13 9 ×11	**14** 10 × 5	**15** 6 ×10	**16** 4 × 6	**17** 9 × 5	**18** 11 ×10	**19** 1 × 7	**20** 4 × 8
21 9 × 1	**22** 8 × 9	**23** 6 × 2	**24** 9 × 6	**25** 5 ×11	**26** 12 × 9	**27** 12 × 6	**28** 12 × 7
29 5 × 8	**30** 11 × 7	**31** 6 × 1	**32** 12 ×12	**33** 2 × 6	**34** 9 ×12	**35** 12 × 4	**36** 11 × 8
37 10 × 2	**38** 9 ×11	**39** 7 × 5	**40** 11 × 8	**41** 4 ×10	**42** 10 × 8	**43** 12 × 0	**44** 10 × 1
45 6 ×12	**46** 0 × 6	**47** 5 × 7	**48** 10 × 9	**49** 3 ×11	**50** 10 × 7	**51** 12 × 8	**52** 11 × 1
53 9 × 0	**54** 4 ×11	**55** 4 ×12	**56** 0 × 9	**57** 11 ×10	**58** 2 ×11	**59** 3 × 9	**60** 7 × 6

© 2024 Teach Elementals, LLC

Name: _____ Date: _____

SCORE

1
11
× 7

2
4
× 7

3
7
×11

4
8
× 9

5
7
× 9

6
10
×12

7
7
× 7

8
11
×11

9
5
×12

10
8
× 2

11
7
× 6

12
8
×12

13
8
× 1

14
7
× 3

15
0
×12

16
8
× 5

17
12
× 7

18
10
×12

19
12
× 9

20
6
×11

21
9
× 7

22
11
× 3

23
8
×10

24
6
×10

25
10
× 7

26
10
× 4

27
10
× 3

28
7
× 8

29
3
×10

30
10
× 6

31
6
× 7

32
6
× 9

33
7
× 9

34
0
× 8

35
9
×10

36
1
×10

37
8
× 7

38
12
× 3

39
2
× 7

40
2
× 6

41
12
×10

42
6
× 9

43
11
× 9

44
1
×11

45
7
×10

46
7
× 2

47
8
× 7

48
12
×11

49
8
× 6

50
6
× 7

51
6
×12

52
7
× 0

53
6
× 3

54
1
× 8

55
2
× 9

56
6
× 0

57
10
×10

58
11
×12

59
3
×12

60
7
×12

© 2024 Teach Elementals, LLC

Name: _____

Date: _____

SCORE

1
$$\begin{array}{r} 5 \\ \times 11 \\ \hline \end{array}$$

2
$$\begin{array}{r} 9 \\ \times 8 \\ \hline \end{array}$$

3
$$\begin{array}{r} 2 \\ \times 8 \\ \hline \end{array}$$

4
$$\begin{array}{r} 4 \\ \times 6 \\ \hline \end{array}$$

5
$$\begin{array}{r} 10 \\ \times 11 \\ \hline \end{array}$$

6
$$\begin{array}{r} 9 \\ \times 6 \\ \hline \end{array}$$

7
$$\begin{array}{r} 8 \\ \times 2 \\ \hline \end{array}$$

8
$$\begin{array}{r} 12 \\ \times 12 \\ \hline \end{array}$$

9
$$\begin{array}{r} 2 \\ \times 12 \\ \hline \end{array}$$

10
$$\begin{array}{r} 11 \\ \times 12 \\ \hline \end{array}$$

11
$$\begin{array}{r} 0 \\ \times 9 \\ \hline \end{array}$$

12
$$\begin{array}{r} 1 \\ \times 7 \\ \hline \end{array}$$

13
$$\begin{array}{r} 6 \\ \times 6 \\ \hline \end{array}$$

14
$$\begin{array}{r} 0 \\ \times 10 \\ \hline \end{array}$$

15
$$\begin{array}{r} 11 \\ \times 6 \\ \hline \end{array}$$

16
$$\begin{array}{r} 12 \\ \times 8 \\ \hline \end{array}$$

17
$$\begin{array}{r} 11 \\ \times 6 \\ \hline \end{array}$$

18
$$\begin{array}{r} 1 \\ \times 10 \\ \hline \end{array}$$

19
$$\begin{array}{r} 9 \\ \times 11 \\ \hline \end{array}$$

20
$$\begin{array}{r} 11 \\ \times 8 \\ \hline \end{array}$$

21
$$\begin{array}{r} 7 \\ \times 9 \\ \hline \end{array}$$

22
$$\begin{array}{r} 10 \\ \times 9 \\ \hline \end{array}$$

23
$$\begin{array}{r} 5 \\ \times 7 \\ \hline \end{array}$$

24
$$\begin{array}{r} 9 \\ \times 12 \\ \hline \end{array}$$

25
$$\begin{array}{r} 0 \\ \times 11 \\ \hline \end{array}$$

26
$$\begin{array}{r} 6 \\ \times 7 \\ \hline \end{array}$$

27
$$\begin{array}{r} 9 \\ \times 2 \\ \hline \end{array}$$

28
$$\begin{array}{r} 11 \\ \times 8 \\ \hline \end{array}$$

29
$$\begin{array}{r} 8 \\ \times 7 \\ \hline \end{array}$$

30
$$\begin{array}{r} 10 \\ \times 7 \\ \hline \end{array}$$

31
$$\begin{array}{r} 7 \\ \times 8 \\ \hline \end{array}$$

32
$$\begin{array}{r} 12 \\ \times 9 \\ \hline \end{array}$$

33
$$\begin{array}{r} 7 \\ \times 11 \\ \hline \end{array}$$

34
$$\begin{array}{r} 12 \\ \times 0 \\ \hline \end{array}$$

35
$$\begin{array}{r} 6 \\ \times 12 \\ \hline \end{array}$$

36
$$\begin{array}{r} 7 \\ \times 0 \\ \hline \end{array}$$

37
$$\begin{array}{r} 4 \\ \times 10 \\ \hline \end{array}$$

38
$$\begin{array}{r} 7 \\ \times 9 \\ \hline \end{array}$$

39
$$\begin{array}{r} 12 \\ \times 1 \\ \hline \end{array}$$

40
$$\begin{array}{r} 7 \\ \times 3 \\ \hline \end{array}$$

41
$$\begin{array}{r} 10 \\ \times 1 \\ \hline \end{array}$$

42
$$\begin{array}{r} 9 \\ \times 4 \\ \hline \end{array}$$

43
$$\begin{array}{r} 12 \\ \times 10 \\ \hline \end{array}$$

44
$$\begin{array}{r} 11 \\ \times 1 \\ \hline \end{array}$$

45
$$\begin{array}{r} 2 \\ \times 6 \\ \hline \end{array}$$

46
$$\begin{array}{r} 7 \\ \times 7 \\ \hline \end{array}$$

47
$$\begin{array}{r} 8 \\ \times 9 \\ \hline \end{array}$$

48
$$\begin{array}{r} 3 \\ \times 10 \\ \hline \end{array}$$

49
$$\begin{array}{r} 6 \\ \times 0 \\ \hline \end{array}$$

50
$$\begin{array}{r} 10 \\ \times 6 \\ \hline \end{array}$$

51
$$\begin{array}{r} 6 \\ \times 12 \\ \hline \end{array}$$

52
$$\begin{array}{r} 8 \\ \times 5 \\ \hline \end{array}$$

53
$$\begin{array}{r} 9 \\ \times 10 \\ \hline \end{array}$$

54
$$\begin{array}{r} 4 \\ \times 12 \\ \hline \end{array}$$

55
$$\begin{array}{r} 12 \\ \times 3 \\ \hline \end{array}$$

56
$$\begin{array}{r} 7 \\ \times 4 \\ \hline \end{array}$$

57
$$\begin{array}{r} 6 \\ \times 8 \\ \hline \end{array}$$

58
$$\begin{array}{r} 12 \\ \times 5 \\ \hline \end{array}$$

59
$$\begin{array}{r} 7 \\ \times 6 \\ \hline \end{array}$$

60
$$\begin{array}{r} 10 \\ \times 9 \\ \hline \end{array}$$

Skill Support: Multiplication Facts 0-12

© 2024 Teach Elementals, LLC

Name: _____   Date: _____

SCORE

1
10
×10

2
12
×10

3
8
×11

4
12
× 2

5
8
× 6

6
12
× 6

7
1
× 6

8
7
×11

9
3
×12

10
2
× 9

11
10
× 7

12
11
×12

13
6
× 1

14
6
×10

15
6
×10

16
10
× 6

17
1
×12

18
7
× 1

19
10
× 3

20
12
× 7

21
9
× 1

22
0
× 6

23
8
×12

24
1
× 8

25
1
× 9

26
6
×11

27
9
×10

28
11
× 2

29
9
×12

30
8
×10

31
7
×12

32
11
×10

33
9
× 7

34
0
× 8

35
5
× 8

36
8
×10

37
6
× 3

38
11
× 4

39
10
× 5

40
9
× 5

41
3
×11

42
11
× 9

43
9
× 3

44
3
× 6

45
12
× 7

46
12
× 4

47
10
× 8

48
6
× 9

49
5
×10

50
9
× 8

51
12
× 8

52
8
×11

53
6
× 8

54
11
× 7

55
7
× 8

56
10
× 0

57
9
× 9

58
7
× 5

59
4
× 8

60
9
× 0

© 2024 Teach Elementals, LLC

SKILL SUPPORT

Name: _____

Date: _____

SCORE

1
1
×11

2
5
×12

3
8
× 4

4
10
× 2

5
7
×12

6
7
× 6

7
9
× 0

8
6
× 8

9
8
× 8

10
3
× 7

11
0
×12

12
4
× 9

13
11
× 3

14
8
× 6

15
3
× 8

16
8
× 0

17
2
×10

18
8
× 1

19
12
× 6

20
2
×11

21
10
×12

22
4
×11

23
5
× 9

24
12
×11

25
8
× 9

26
10
× 4

27
6
×11

28
7
×10

29
7
×10

30
6
× 5

31
5
× 6

32
10
×11

33
12
× 9

34
8
×12

35
4
× 8

36
7
× 5

37
11
× 5

38
6
× 7

39
7
× 4

40
8
× 3

41
11
× 7

42
10
× 8

43
8
× 7

44
9
× 7

45
2
× 7

46
6
× 2

47
6
× 4

48
3
× 9

49
6
× 9

50
9
×11

51
10
×12

52
12
×11

53
11
× 9

54
11
× 0

55
4
× 7

56
7
× 2

57
9
× 6

58
0
× 7

59
11
×11

60
11
×10

© 2024 Teach Elementals, LLC

Name: _____   Date: _____

SCORE

1
2
× 2

2
4
× 0

3
10
× 2

4
8
× 2

5
12
× 2

6
12
× 6

7
10
× 8

8
4
× 6

9
4
× 8

10
0
×10

11
10
× 6

12
8
× 8

13
2
× 6

14
4
×12

15
8
×10

16
6
×10

17
4
× 2

18
12
× 4

19
2
× 8

20
10
× 4

21
8
× 4

22
12
×10

23
10
×12

24
6
× 4

25
8
×12

26
6
× 2

27
12
× 8

28
10
×10

29
4
×10

30
6
×12

31
2
×12

32
6
× 6

33
6
× 8

34
8
× 6

35
2
× 4

36
12
×12

37
0
×12

38
2
×10

39
6
× 8

40
8
× 6

41
4
× 8

42
4
× 4

43
10
× 8

44
6
× 6

45
12
× 2

46
10
×10

47
10
× 4

48
2
×12

49
4
× 2

50
6
×10

51
2
× 6

52
2
× 2

53
12
× 4

54
10
× 0

55
4
× 6

56
10
×12

57
12
×10

58
8
× 4

59
8
× 8

60
6
× 4

© 2024 Teach Elementals, LLC

MIXED REVIEW: EVENS ONLY, SET B

Name: _____

Date: _____

SCORE

1 6 × 6	**2** 6 × 4	**3** 8 × 6	**4** 0 × 4
5 10 ×10	**6** 12 × 8	**7** 10 × 6	**8** 6 ×10
9 2 × 6	**10** 6 × 2	**11** 6 ×12	**12** 12 × 0
13 12 ×12	**14** 4 × 2	**15** 10 × 2	**16** 4 ×10
17 4 × 8	**18** 10 ×12	**19** 6 × 8	**20** 8 × 8
21 8 × 2	**22** 10 × 8	**23** 4 ×12	**24** 2 ×10
25 8 × 0	**26** 12 × 6	**27** 2 × 8	**28** 10 × 4
29 12 × 4	**30** 12 × 2	**31** 2 ×12	**32** 2 × 2
33 8 ×10	**34** 2 × 4	**35** 8 ×12	**36** 4 × 6
37 12 × 8	**38** 10 × 4	**39** 4 × 6	**40** 4 × 2
41 4 × 8	**42** 2 ×12	**43** 6 × 2	**44** 8 ×12
45 6 ×10	**46** 10 × 2	**47** 2 × 2	**48** 6 × 8
49 2 × 6	**50** 4 ×12	**51** 8 × 8	**52** 6 × 6
53 4 × 4	**54** 10 × 0	**55** 12 ×10	**56** 12 × 4
57 4 ×10	**58** 2 × 4	**59** 10 × 6	**60** 2 × 8

© 2024 Teach Elementals, LLC

Name: _____ Date: _____

SCORE

1
10
×10

2
2
× 0

3
2
×10

4
2
× 2

5
12
× 4

6
6
× 6

7
6
×10

8
12
× 8

9
8
×10

10
0
× 6

11
4
× 8

12
4
× 4

13
6
× 4

14
8
× 6

15
6
× 2

16
2
× 8

17
10
× 8

18
10
× 6

19
10
× 4

20
8
× 4

21
4
× 2

22
12
×10

23
12
×12

24
4
×10

25
10
×12

26
8
×12

27
2
× 6

28
12
× 2

29
6
×12

30
6
× 8

31
4
×12

32
12
× 6

33
2
× 4

34
8
× 2

35
10
× 2

36
8
× 8

37
0
×10

38
8
×12

39
2
×10

40
2
×12

41
4
×12

42
4
× 6

43
12
× 8

44
6
×10

45
10
× 8

46
4
× 8

47
6
×12

48
10
×10

49
8
×10

50
2
× 2

51
2
× 8

52
6
× 2

53
4
× 2

54
6
× 0

55
8
× 8

56
12
× 2

57
4
× 4

58
8
× 4

59
10
× 4

60
12
× 4

© 2024 Teach Elementals, LLC

Skill Support: Multiplication Facts 0-12   55

MIXED REVIEW: EVENS ONLY, SET D

Name: _____

Date: _____

SCORE

1 10 × 2	**2** 2 ×10	**3** 8 × 6	**4** 0 × 6

5 4 ×12	**6** 8 × 8	**7** 4 × 8	**8** 10 × 6	**9** 12 × 2	**10** 4 × 2	**11** 4 × 6	**12** 12 × 0
13 2 × 2	**14** 6 ×10	**15** 6 × 4	**16** 6 × 8	**17** 8 × 2	**18** 12 × 6	**19** 12 ×10	**20** 2 × 4
21 8 ×10	**22** 2 × 8	**23** 12 × 4	**24** 6 ×12	**25** 12 × 0	**26** 2 × 6	**27** 10 × 8	**28** 4 ×10
29 2 ×12	**30** 8 × 4	**31** 6 × 2	**32** 10 ×12	**33** 8 ×12	**34** 10 ×10	**35** 10 × 4	**36** 4 × 4
37 10 ×10	**38** 10 × 2	**39** 6 × 6	**40** 4 × 8	**41** 2 × 4	**42** 12 × 4	**43** 4 × 4	**44** 6 × 4
45 4 ×10	**46** 2 ×12	**47** 12 ×12	**48** 6 × 8	**49** 4 × 2	**50** 4 ×12	**51** 6 ×10	**52** 10 × 6
53 2 × 8	**54** 12 × 0	**55** 8 × 2	**56** 8 ×12	**57** 12 × 2	**58** 8 × 8	**59** 8 × 6	**60** 8 × 4

Skill Support: Multiplication Facts 0-12

© 2024 Teach Elementals, LLC

Name: _____ Date: _____

SCORE

1	**2**	**3**	**4**
7 × 11	1 × 1	11 × 1	1 × 11

5	**6**	**7**	**8**	**9**	**10**	**11**	**12**
1 × 5	3 × 7	1 × 9	9 × 5	11 × 5	7 × 3	5 × 9	9 × 1
13	**14**	**15**	**16**	**17**	**18**	**19**	**20**
7 × 1	7 × 5	11 × 11	9 × 3	5 × 5	11 × 7	11 × 3	3 × 11
21	**22**	**23**	**24**	**25**	**26**	**27**	**28**
11 × 9	9 × 9	9 × 7	1 × 7	5 × 7	9 × 11	7 × 7	5 × 11
29	**30**	**31**	**32**	**33**	**34**	**35**	**36**
5 × 1	3 × 5	7 × 9	3 × 1	5 × 3	3 × 3	1 × 3	3 × 9
37	**38**	**39**	**40**	**41**	**42**	**43**	**44**
5 × 7	7 × 11	5 × 11	3 × 1	9 × 7	11 × 5	3 × 5	1 × 5
45	**46**	**47**	**48**	**49**	**50**	**51**	**52**
11 × 1	9 × 11	1 × 11	11 × 7	1 × 7	7 × 3	5 × 9	3 × 3
53	**54**	**55**	**56**	**57**	**58**	**59**	**60**
11 × 9	7 × 1	3 × 9	5 × 3	9 × 1	1 × 1	1 × 9	1 × 3

MATH SKILL SUPPORT

MIXED REVIEW: ODDS ONLY, SET B

Name: _____

Date: _____

SCORE

1 11 × 3	**2** 3 × 3	**3** 5 × 3	**4** 5 × 9
5 3 × 1	**6** 7 × 5	**7** 9 ×11	**8** 9 × 7
9 7 ×11	**10** 9 × 5	**11** 3 × 5	**12** 3 ×11
13 11 × 1	**14** 5 × 5	**15** 7 × 1	**16** 1 × 3
17 11 × 5	**18** 5 ×11	**19** 3 × 9	**20** 7 × 3
21 11 ×11	**22** 5 × 1	**23** 7 × 9	**24** 9 × 3
25 9 × 9	**26** 9 × 1	**27** 5 × 7	**28** 1 × 7
29 1 ×11	**30** 11 × 7	**31** 1 × 5	**32** 3 × 7
33 7 × 7	**34** 1 × 9	**35** 11 × 9	**36** 1 × 1
37 7 × 3	**38** 7 × 7	**39** 1 × 3	**40** 9 × 1
41 9 × 5	**42** 3 × 5	**43** 7 ×11	**44** 11 × 7
45 9 ×11	**46** 1 × 7	**47** 5 ×11	**48** 11 × 5
49 1 × 9	**50** 9 × 3	**51** 9 × 7	**52** 5 × 9
53 11 × 1	**54** 3 × 1	**55** 11 × 9	**56** 3 × 9
57 5 × 3	**58** 5 × 5	**59** 11 ×11	**60** 1 × 5

© 2024 Teach Elementals, LLC

Name: _____

Date: _____

SCORE

1 1 × 5	**2** 7 × 3	**3** 1 × 3	**4** 3 ×11
5 5 × 9	**6** 5 × 7	**7** 1 × 9	**8** 9 ×11
9 1 × 1	**10** 5 ×11	**11** 5 × 5	**12** 9 × 1
13 3 × 5	**14** 11 × 5	**15** 9 × 9	**16** 11 ×11
17 7 × 1	**18** 7 × 9	**19** 3 × 1	**20** 3 × 3
21 3 × 7	**22** 11 × 7	**23** 7 ×11	**24** 9 × 3
25 5 × 1	**26** 7 × 5	**27** 1 × 7	**28** 9 × 5
29 5 × 3	**30** 3 × 9	**31** 9 × 7	**32** 11 × 9
33 11 × 3	**34** 11 × 1	**35** 1 ×11	**36** 7 × 7
37 1 × 5	**38** 1 × 9	**39** 3 ×11	**40** 5 × 5
41 9 × 7	**42** 11 × 9	**43** 9 × 9	**44** 5 × 3
45 11 × 3	**46** 3 × 3	**47** 7 × 9	**48** 7 × 5
49 11 × 5	**50** 5 × 1	**51** 1 × 3	**52** 5 ×11
53 5 × 9	**54** 9 × 3	**55** 9 × 5	**56** 11 × 7
57 3 × 1	**58** 7 ×11	**59** 11 × 1	**60** 1 × 1

Name: _____   Date: _____

SCORE

1	2	3	4
5 × 7	7 ×11	3 × 1	9 × 3

5	6	7	8	9	10	11	12
1 × 9	9 × 9	11 ×11	1 × 7	11 × 9	9 × 1	11 × 3	3 × 5

13	14	15	16	17	18	19	20
9 ×11	3 ×11	11 × 5	7 × 9	1 × 1	11 × 7	3 × 3	5 × 1

21	22	23	24	25	26	27	28
1 × 3	5 × 3	1 ×11	7 × 3	9 × 5	5 ×11	9 × 7	3 × 7

29	30	31	32	33	34	35	36
5 × 5	7 × 1	5 × 9	7 × 7	7 × 5	3 × 9	1 × 5	11 × 1

37	38	39	40	41	42	43	44
3 × 5	5 × 7	1 × 3	7 × 1	3 × 3	5 ×11	7 × 9	5 × 3

45	46	47	48	49	50	51	52
9 × 7	7 × 3	7 ×11	5 × 9	3 × 9	11 × 9	9 × 3	9 × 5

53	54	55	56	57	58	59	60
7 × 7	1 × 5	11 ×11	1 × 1	1 ×11	1 × 9	3 × 1	3 ×11

© 2024 Teach Elementals, LLC

SCORE

1
4
× 1

2
8
× 1

3
8
× 3

4
8
× 2

5
2
× 9

6
8
× 4

7
8
× 8

8
2
× 6

9
2
× 2

10
4
× 4

11
2
×12

12
4
× 5

13
4
× 6

14
2
× 8

15
2
× 4

16
4
×10

17
4
× 2

18
2
× 0

19
8
×12

20
8
× 0

21
4
× 9

22
2
× 3

23
2
× 5

24
2
×11

25
2
×10

26
4
× 0

27
8
× 9

28
8
×10

29
4
×11

30
4
× 3

31
4
× 7

32
8
×11

33
8
× 7

34
4
× 8

35
8
× 5

36
2
× 1

37
4
×12

38
2
× 7

39
8
× 6

40
9
× 8

41
0
× 2

42
11
× 2

43
12
× 2

44
0
× 8

45
10
× 4

46
9
× 2

47
9
× 4

48
8
× 2

49
6
× 2

50
4
× 2

51
5
× 2

52
1
× 4

53
5
× 8

54
2
× 4

55
11
× 8

56
11
× 4

57
12
× 8

58
0
× 4

59
6
× 4

60
8
× 4

© 2024 Teach Elementals, LLC

MIXED REVIEW, 2, 4, AND 8, SET B

Name: _____   Date: _____

SCORE

1 2 × 2	**2** 2 ×10	**3** 8 × 7	**4** 4 × 4
5 4 × 5	**6** 4 × 0	**7** 2 × 9	**8** 2 × 7
9 4 ×11	**10** 2 × 8	**11** 4 × 2	**12** 2 ×12
13 4 × 6	**14** 2 ×11	**15** 4 × 3	**16** 2 × 4
17 2 × 3	**18** 8 × 8	**19** 8 ×11	**20** 8 ×10
21 4 × 7	**22** 8 × 4	**23** 8 × 9	**24** 8 × 0
25 2 × 1	**26** 2 × 6	**27** 8 × 3	**28** 4 ×12
29 8 × 2	**30** 4 × 9	**31** 8 × 6	**32** 2 × 0
33 4 ×10	**34** 2 × 5	**35** 4 × 8	**36** 8 × 5
37 8 × 1	**38** 4 × 1	**39** 8 ×12	**40** 0 × 4
41 11 × 8	**42** 0 × 8	**43** 2 × 4	**44** 12 × 8
45 4 × 2	**46** 3 × 8	**47** 1 × 2	**48** 6 × 2
49 1 × 8	**50** 10 × 2	**51** 2 × 8	**52** 3 × 2
53 3 × 4	**54** 6 × 4	**55** 7 × 4	**56** 7 × 8
57 4 × 8	**58** 12 × 4	**59** 1 × 4	**60** 10 × 4

© 2024 Teach Elementals, LLC

Name: _____   Date: _____

SCORE

#		#		#		#	
1	8 × 0	2	2 × 8	3	8 × 12	4	4 × 8

#		#		#		#	
5	2 × 3	6	4 × 7	7	8 × 7	8	4 × 12
9	8 × 11	10	4 × 2	11	4 × 4	12	2 × 2
13	2 × 12	14	2 × 4	15	4 × 10	16	4 × 11
17	4 × 5	18	2 × 1	19	8 × 9	20	8 × 5
21	2 × 6	22	8 × 10	23	8 × 4	24	2 × 10
25	4 × 0	26	2 × 7	27	8 × 1	28	8 × 3
29	4 × 1	30	8 × 8	31	2 × 0	32	4 × 9
33	2 × 5	34	2 × 9	35	4 × 3	36	4 × 6
37	2 × 11	38	8 × 2	39	8 × 6	40	1 × 4
41	0 × 4	42	9 × 8	43	5 × 2	44	2 × 8
45	6 × 2	46	11 × 2	47	0 × 8	48	1 × 2
49	6 × 8	50	12 × 4	51	4 × 2	52	7 × 8
53	7 × 4	54	6 × 4	55	12 × 8	56	10 × 4
57	11 × 8	58	10 × 8	59	11 × 4	60	8 × 4

© 2024 Teach Elementals, LLC

Name: _____   Date: _____

SCORE

1 $\begin{array}{r} 9 \\ \times\ 8 \\ \hline \end{array}$	**2** $\begin{array}{r} 6 \\ \times\ 4 \\ \hline \end{array}$	**3** $\begin{array}{r} 3 \\ \times 10 \\ \hline \end{array}$	**4** $\begin{array}{r} 6 \\ \times 10 \\ \hline \end{array}$

5 $\begin{array}{r} 3 \\ \times 12 \\ \hline \end{array}$	**6** $\begin{array}{r} 6 \\ \times\ 2 \\ \hline \end{array}$	**7** $\begin{array}{r} 6 \\ \times\ 7 \\ \hline \end{array}$	**8** $\begin{array}{r} 9 \\ \times\ 6 \\ \hline \end{array}$	**9** $\begin{array}{r} 3 \\ \times\ 4 \\ \hline \end{array}$	**10** $\begin{array}{r} 6 \\ \times 11 \\ \hline \end{array}$	**11** $\begin{array}{r} 9 \\ \times\ 9 \\ \hline \end{array}$	**12** $\begin{array}{r} 9 \\ \times\ 7 \\ \hline \end{array}$
13 $\begin{array}{r} 3 \\ \times 11 \\ \hline \end{array}$	**14** $\begin{array}{r} 3 \\ \times\ 6 \\ \hline \end{array}$	**15** $\begin{array}{r} 9 \\ \times\ 0 \\ \hline \end{array}$	**16** $\begin{array}{r} 3 \\ \times\ 0 \\ \hline \end{array}$	**17** $\begin{array}{r} 6 \\ \times\ 9 \\ \hline \end{array}$	**18** $\begin{array}{r} 3 \\ \times\ 2 \\ \hline \end{array}$	**19** $\begin{array}{r} 3 \\ \times\ 3 \\ \hline \end{array}$	**20** $\begin{array}{r} 6 \\ \times\ 5 \\ \hline \end{array}$
21 $\begin{array}{r} 9 \\ \times\ 4 \\ \hline \end{array}$	**22** $\begin{array}{r} 6 \\ \times\ 3 \\ \hline \end{array}$	**23** $\begin{array}{r} 3 \\ \times\ 9 \\ \hline \end{array}$	**24** $\begin{array}{r} 6 \\ \times\ 0 \\ \hline \end{array}$	**25** $\begin{array}{r} 6 \\ \times\ 1 \\ \hline \end{array}$	**26** $\begin{array}{r} 6 \\ \times\ 8 \\ \hline \end{array}$	**27** $\begin{array}{r} 9 \\ \times 10 \\ \hline \end{array}$	**28** $\begin{array}{r} 9 \\ \times\ 1 \\ \hline \end{array}$
29 $\begin{array}{r} 9 \\ \times 12 \\ \hline \end{array}$	**30** $\begin{array}{r} 6 \\ \times 12 \\ \hline \end{array}$	**31** $\begin{array}{r} 9 \\ \times 11 \\ \hline \end{array}$	**32** $\begin{array}{r} 3 \\ \times\ 8 \\ \hline \end{array}$	**33** $\begin{array}{r} 9 \\ \times\ 5 \\ \hline \end{array}$	**34** $\begin{array}{r} 9 \\ \times\ 2 \\ \hline \end{array}$	**35** $\begin{array}{r} 3 \\ \times\ 5 \\ \hline \end{array}$	**36** $\begin{array}{r} 9 \\ \times\ 3 \\ \hline \end{array}$
37 $\begin{array}{r} 3 \\ \times\ 7 \\ \hline \end{array}$	**38** $\begin{array}{r} 3 \\ \times\ 1 \\ \hline \end{array}$	**39** $\begin{array}{r} 6 \\ \times\ 6 \\ \hline \end{array}$	**40** $\begin{array}{r} 1 \\ \times\ 9 \\ \hline \end{array}$	**41** $\begin{array}{r} 6 \\ \times\ 3 \\ \hline \end{array}$	**42** $\begin{array}{r} 9 \\ \times\ 3 \\ \hline \end{array}$	**43** $\begin{array}{r} 6 \\ \times\ 9 \\ \hline \end{array}$	**44** $\begin{array}{r} 11 \\ \times\ 6 \\ \hline \end{array}$
45 $\begin{array}{r} 2 \\ \times\ 3 \\ \hline \end{array}$	**46** $\begin{array}{r} 8 \\ \times\ 6 \\ \hline \end{array}$	**47** $\begin{array}{r} 9 \\ \times\ 6 \\ \hline \end{array}$	**48** $\begin{array}{r} 11 \\ \times\ 9 \\ \hline \end{array}$	**49** $\begin{array}{r} 10 \\ \times\ 9 \\ \hline \end{array}$	**50** $\begin{array}{r} 7 \\ \times\ 9 \\ \hline \end{array}$	**51** $\begin{array}{r} 10 \\ \times\ 6 \\ \hline \end{array}$	**52** $\begin{array}{r} 8 \\ \times\ 3 \\ \hline \end{array}$
53 $\begin{array}{r} 5 \\ \times\ 9 \\ \hline \end{array}$	**54** $\begin{array}{r} 1 \\ \times\ 6 \\ \hline \end{array}$	**55** $\begin{array}{r} 10 \\ \times\ 3 \\ \hline \end{array}$	**56** $\begin{array}{r} 0 \\ \times\ 6 \\ \hline \end{array}$	**57** $\begin{array}{r} 2 \\ \times\ 6 \\ \hline \end{array}$	**58** $\begin{array}{r} 5 \\ \times\ 6 \\ \hline \end{array}$	**59** $\begin{array}{r} 8 \\ \times\ 9 \\ \hline \end{array}$	**60** $\begin{array}{r} 7 \\ \times\ 3 \\ \hline \end{array}$

© 2024 Teach Elementals, LLC

Name: _____   Date: _____

SCORE

1 9×1	**2** 3×9	**3** 9×4	**4** 6×4

5 3×6	**6** 6×7	**7** 3×1	**8** 9×8	**9** 9×12	**10** 9×5	**11** 9×0	**12** 3×11

13 9×11	**14** 3×5	**15** 6×10	**16** 6×3	**17** 9×7	**18** 9×9	**19** 6×2	**20** 9×3

21 9×6	**22** 6×9	**23** 6×5	**24** 6×12	**25** 3×0	**26** 3×2	**27** 9×10	**28** 3×4

29 3×8	**30** 9×2	**31** 3×10	**32** 6×1	**33** 6×11	**34** 3×12	**35** 6×6	**36** 3×7

37 6×8	**38** 6×0	**39** 3×3	**40** 9×6	**41** 0×3	**42** 5×6	**43** 10×6	**44** 2×6

45 1×6	**46** 6×3	**47** 4×3	**48** 3×9	**49** 6×9	**50** 1×3	**51** 4×9	**52** 2×9

53 12×3	**54** 1×9	**55** 11×9	**56** 8×6	**57** 11×3	**58** 9×3	**59** 7×6	**60** 2×3

© 2024 Teach Elementals, LLC

Name: _____ Date: _____

SCORE

1	2	3	4
9 × 1	9 × 5	9 × 11	3 × 3

5	6	7	8	9	10	11	12
3 × 12	3 × 10	9 × 12	6 × 0	3 × 5	3 × 7	9 × 8	9 × 9

13	14	15	16	17	18	19	20
9 × 2	3 × 8	6 × 4	3 × 6	6 × 2	3 × 0	9 × 7	6 × 6

21	22	23	24	25	26	27	28
3 × 2	9 × 10	3 × 9	9 × 3	9 × 4	6 × 7	6 × 3	6 × 10

29	30	31	32	33	34	35	36
6 × 9	6 × 12	6 × 8	3 × 4	9 × 6	6 × 1	3 × 11	9 × 0

37	38	39	40	41	42	43	44
6 × 5	3 × 1	6 × 11	4 × 6	1 × 9	6 × 9	0 × 3	9 × 3

45	46	47	48	49	50	51	52
1 × 3	12 × 9	12 × 6	7 × 3	2 × 9	10 × 6	11 × 3	1 × 6

53	54	55	56	57	58	59	60
2 × 3	10 × 3	5 × 3	8 × 6	9 × 6	6 × 3	4 × 3	5 × 9

© 2024 Teach Elementals, LLC

Name: _____   Date: _____

SCORE

1	2	3	4
10 × 2	7 × 5	5 ×12	11 ×10

5	6	7	8	9	10	11	12
3 ×10	10 × 1	5 × 3	5 × 5	8 ×10	4 ×10	10 × 5	5 × 8

13	14	15	16	17	18	19	20
10 ×12	2 ×10	9 ×10	10 × 7	9 × 5	12 ×10	12 × 5	10 × 8

21	22	23	24	25	26	27	28
1 × 5	5 ×10	5 × 0	5 × 2	8 × 5	10 × 5	7 ×10	5 ×11

29	30	31	32	33	34	35	36
0 × 5	10 × 9	11 × 5	4 × 5	10 × 6	5 × 9	3 × 5	10 ×10

37	38	39	40	41	42	43	44
10 × 4	1 ×10	0 ×10	5 ×10	6 ×10	10 ×11	6 × 5	5 × 1

45	46	47	48	49	50	51	52
5 × 4	5 × 6	10 × 3	5 × 7	10 × 0	2 × 5	8 × 5	11 ×10

53	54	55	56	57	58	59	60
5 × 5	10 ×11	10 × 9	0 × 5	10 ×12	10 × 5	10 × 6	10 × 0

MIXED REVIEW: 5 AND 10, SET B

Name: _____

Date: _____

SCORE

1 $\begin{array}{r} 1 \\ \times\ 5 \\ \hline \end{array}$	**2** $\begin{array}{r} 9 \\ \times 10 \\ \hline \end{array}$	**3** $\begin{array}{r} 11 \\ \times\ 5 \\ \hline \end{array}$	**4** $\begin{array}{r} 5 \\ \times\ 4 \\ \hline \end{array}$

5 $\begin{array}{r} 10 \\ \times\ 9 \\ \hline \end{array}$	**6** $\begin{array}{r} 10 \\ \times\ 4 \\ \hline \end{array}$	**7** $\begin{array}{r} 4 \\ \times\ 5 \\ \hline \end{array}$	**8** $\begin{array}{r} 10 \\ \times\ 5 \\ \hline \end{array}$	**9** $\begin{array}{r} 5 \\ \times 10 \\ \hline \end{array}$	**10** $\begin{array}{r} 5 \\ \times\ 9 \\ \hline \end{array}$	**11** $\begin{array}{r} 1 \\ \times 10 \\ \hline \end{array}$	**12** $\begin{array}{r} 5 \\ \times\ 2 \\ \hline \end{array}$
13 $\begin{array}{r} 3 \\ \times\ 5 \\ \hline \end{array}$	**14** $\begin{array}{r} 5 \\ \times\ 3 \\ \hline \end{array}$	**15** $\begin{array}{r} 10 \\ \times 10 \\ \hline \end{array}$	**16** $\begin{array}{r} 10 \\ \times\ 5 \\ \hline \end{array}$	**17** $\begin{array}{r} 8 \\ \times\ 5 \\ \hline \end{array}$	**18** $\begin{array}{r} 5 \\ \times\ 7 \\ \hline \end{array}$	**19** $\begin{array}{r} 8 \\ \times 10 \\ \hline \end{array}$	**20** $\begin{array}{r} 7 \\ \times 10 \\ \hline \end{array}$
21 $\begin{array}{r} 5 \\ \times\ 5 \\ \hline \end{array}$	**22** $\begin{array}{r} 11 \\ \times 10 \\ \hline \end{array}$	**23** $\begin{array}{r} 10 \\ \times\ 8 \\ \hline \end{array}$	**24** $\begin{array}{r} 10 \\ \times 12 \\ \hline \end{array}$	**25** $\begin{array}{r} 12 \\ \times\ 5 \\ \hline \end{array}$	**26** $\begin{array}{r} 10 \\ \times\ 6 \\ \hline \end{array}$	**27** $\begin{array}{r} 10 \\ \times\ 0 \\ \hline \end{array}$	**28** $\begin{array}{r} 5 \\ \times 11 \\ \hline \end{array}$
29 $\begin{array}{r} 7 \\ \times\ 5 \\ \hline \end{array}$	**30** $\begin{array}{r} 5 \\ \times 12 \\ \hline \end{array}$	**31** $\begin{array}{r} 5 \\ \times\ 1 \\ \hline \end{array}$	**32** $\begin{array}{r} 0 \\ \times 10 \\ \hline \end{array}$	**33** $\begin{array}{r} 5 \\ \times 10 \\ \hline \end{array}$	**34** $\begin{array}{r} 6 \\ \times 10 \\ \hline \end{array}$	**35** $\begin{array}{r} 5 \\ \times\ 0 \\ \hline \end{array}$	**36** $\begin{array}{r} 10 \\ \times 11 \\ \hline \end{array}$
37 $\begin{array}{r} 12 \\ \times 10 \\ \hline \end{array}$	**38** $\begin{array}{r} 9 \\ \times\ 5 \\ \hline \end{array}$	**39** $\begin{array}{r} 4 \\ \times 10 \\ \hline \end{array}$	**40** $\begin{array}{r} 3 \\ \times 10 \\ \hline \end{array}$	**41** $\begin{array}{r} 6 \\ \times\ 5 \\ \hline \end{array}$	**42** $\begin{array}{r} 0 \\ \times\ 5 \\ \hline \end{array}$	**43** $\begin{array}{r} 10 \\ \times\ 7 \\ \hline \end{array}$	**44** $\begin{array}{r} 10 \\ \times\ 2 \\ \hline \end{array}$
45 $\begin{array}{r} 10 \\ \times\ 3 \\ \hline \end{array}$	**46** $\begin{array}{r} 2 \\ \times\ 5 \\ \hline \end{array}$	**47** $\begin{array}{r} 2 \\ \times 10 \\ \hline \end{array}$	**48** $\begin{array}{r} 10 \\ \times\ 1 \\ \hline \end{array}$	**49** $\begin{array}{r} 5 \\ \times\ 6 \\ \hline \end{array}$	**50** $\begin{array}{r} 5 \\ \times\ 8 \\ \hline \end{array}$	**51** $\begin{array}{r} 12 \\ \times 10 \\ \hline \end{array}$	**52** $\begin{array}{r} 5 \\ \times\ 9 \\ \hline \end{array}$
53 $\begin{array}{r} 10 \\ \times\ 2 \\ \hline \end{array}$	**54** $\begin{array}{r} 5 \\ \times\ 3 \\ \hline \end{array}$	**55** $\begin{array}{r} 5 \\ \times\ 5 \\ \hline \end{array}$	**56** $\begin{array}{r} 11 \\ \times\ 5 \\ \hline \end{array}$	**57** $\begin{array}{r} 10 \\ \times\ 6 \\ \hline \end{array}$	**58** $\begin{array}{r} 5 \\ \times\ 2 \\ \hline \end{array}$	**59** $\begin{array}{r} 10 \\ \times\ 7 \\ \hline \end{array}$	**60** $\begin{array}{r} 1 \\ \times\ 5 \\ \hline \end{array}$

© 2024 Teach Elementals, LLC

Name: _____ Date: _____

SCORE

1
5
× 6

2
5
× 8

3
5
×10

4
0
×10

5
4
×10

6
9
× 5

7
7
× 5

8
5
× 7

9
5
× 0

10
5
× 3

11
11
× 5

12
5
×11

13
10
× 7

14
2
×10

15
5
× 4

16
5
× 9

17
10
×11

18
8
× 5

19
5
× 1

20
5
×10

21
10
× 4

22
5
× 2

23
3
×10

24
10
× 9

25
8
×10

26
10
×10

27
10
× 5

28
1
×10

29
2
× 5

30
3
× 5

31
10
× 5

32
10
× 6

33
10
× 3

34
10
× 8

35
6
×10

36
10
× 2

37
10
× 0

38
11
×10

39
5
×12

40
10
× 1

41
0
× 5

42
10
×12

43
5
× 5

44
9
×10

45
1
× 5

46
7
×10

47
12
×10

48
12
× 5

49
4
× 5

50
6
× 5

51
10
×12

52
5
× 5

53
9
×10

54
10
× 3

55
2
× 5

56
10
× 9

57
5
× 9

58
6
× 5

59
10
× 0

60
10
× 2

© 2024 Teach Elementals, LLC

Name: _____

Date: _____

SCORE

1 $\begin{array}{r} 12 \\ \times\ 4 \\ \hline \end{array}$	**2** $\begin{array}{r} 6 \\ \times\ 7 \\ \hline \end{array}$	**3** $\begin{array}{r} 6 \\ \times\ 0 \\ \hline \end{array}$	**4** $\begin{array}{r} 6 \\ \times 11 \\ \hline \end{array}$

5 $\begin{array}{r} 6 \\ \times 12 \\ \hline \end{array}$	**6** $\begin{array}{r} 9 \\ \times\ 5 \\ \hline \end{array}$	**7** $\begin{array}{r} 6 \\ \times\ 6 \\ \hline \end{array}$	**8** $\begin{array}{r} 9 \\ \times\ 0 \\ \hline \end{array}$	**9** $\begin{array}{r} 12 \\ \times 12 \\ \hline \end{array}$	**10** $\begin{array}{r} 9 \\ \times\ 7 \\ \hline \end{array}$	**11** $\begin{array}{r} 9 \\ \times\ 1 \\ \hline \end{array}$	**12** $\begin{array}{r} 9 \\ \times 10 \\ \hline \end{array}$
13 $\begin{array}{r} 6 \\ \times\ 5 \\ \hline \end{array}$	**14** $\begin{array}{r} 9 \\ \times 12 \\ \hline \end{array}$	**15** $\begin{array}{r} 9 \\ \times 11 \\ \hline \end{array}$	**16** $\begin{array}{r} 9 \\ \times\ 3 \\ \hline \end{array}$	**17** $\begin{array}{r} 12 \\ \times\ 5 \\ \hline \end{array}$	**18** $\begin{array}{r} 9 \\ \times\ 4 \\ \hline \end{array}$	**19** $\begin{array}{r} 12 \\ \times\ 8 \\ \hline \end{array}$	**20** $\begin{array}{r} 12 \\ \times\ 9 \\ \hline \end{array}$
21 $\begin{array}{r} 6 \\ \times 10 \\ \hline \end{array}$	**22** $\begin{array}{r} 12 \\ \times\ 3 \\ \hline \end{array}$	**23** $\begin{array}{r} 12 \\ \times\ 7 \\ \hline \end{array}$	**24** $\begin{array}{r} 12 \\ \times\ 1 \\ \hline \end{array}$	**25** $\begin{array}{r} 6 \\ \times\ 1 \\ \hline \end{array}$	**26** $\begin{array}{r} 6 \\ \times\ 8 \\ \hline \end{array}$	**27** $\begin{array}{r} 12 \\ \times\ 0 \\ \hline \end{array}$	**28** $\begin{array}{r} 6 \\ \times\ 2 \\ \hline \end{array}$
29 $\begin{array}{r} 9 \\ \times\ 6 \\ \hline \end{array}$	**30** $\begin{array}{r} 12 \\ \times 10 \\ \hline \end{array}$	**31** $\begin{array}{r} 6 \\ \times\ 9 \\ \hline \end{array}$	**32** $\begin{array}{r} 6 \\ \times\ 3 \\ \hline \end{array}$	**33** $\begin{array}{r} 12 \\ \times\ 2 \\ \hline \end{array}$	**34** $\begin{array}{r} 9 \\ \times\ 8 \\ \hline \end{array}$	**35** $\begin{array}{r} 12 \\ \times\ 6 \\ \hline \end{array}$	**36** $\begin{array}{r} 9 \\ \times\ 9 \\ \hline \end{array}$
37 $\begin{array}{r} 9 \\ \times\ 2 \\ \hline \end{array}$	**38** $\begin{array}{r} 6 \\ \times\ 4 \\ \hline \end{array}$	**39** $\begin{array}{r} 12 \\ \times 11 \\ \hline \end{array}$	**40** $\begin{array}{r} 7 \\ \times 12 \\ \hline \end{array}$	**41** $\begin{array}{r} 8 \\ \times\ 9 \\ \hline \end{array}$	**42** $\begin{array}{r} 6 \\ \times\ 9 \\ \hline \end{array}$	**43** $\begin{array}{r} 10 \\ \times\ 6 \\ \hline \end{array}$	**44** $\begin{array}{r} 9 \\ \times 12 \\ \hline \end{array}$
45 $\begin{array}{r} 0 \\ \times\ 9 \\ \hline \end{array}$	**46** $\begin{array}{r} 2 \\ \times\ 9 \\ \hline \end{array}$	**47** $\begin{array}{r} 3 \\ \times\ 6 \\ \hline \end{array}$	**48** $\begin{array}{r} 5 \\ \times\ 6 \\ \hline \end{array}$	**49** $\begin{array}{r} 3 \\ \times 12 \\ \hline \end{array}$	**50** $\begin{array}{r} 9 \\ \times\ 6 \\ \hline \end{array}$	**51** $\begin{array}{r} 10 \\ \times\ 9 \\ \hline \end{array}$	**52** $\begin{array}{r} 12 \\ \times\ 6 \\ \hline \end{array}$
53 $\begin{array}{r} 11 \\ \times\ 9 \\ \hline \end{array}$	**54** $\begin{array}{r} 1 \\ \times 12 \\ \hline \end{array}$	**55** $\begin{array}{r} 5 \\ \times\ 9 \\ \hline \end{array}$	**56** $\begin{array}{r} 0 \\ \times 12 \\ \hline \end{array}$	**57** $\begin{array}{r} 7 \\ \times\ 9 \\ \hline \end{array}$	**58** $\begin{array}{r} 12 \\ \times\ 9 \\ \hline \end{array}$	**59** $\begin{array}{r} 8 \\ \times\ 6 \\ \hline \end{array}$	**60** $\begin{array}{r} 4 \\ \times 12 \\ \hline \end{array}$

© 2024 Teach Elementals, LLC

Name: _____   Date: _____

SCORE

#		#		#		#	
1	9 × 3	**2**	9 × 8	**3**	9 × 0	**4**	12 ×10

| **5** | 12 × 6 | **6** | 12 × 4 | **7** | 9 × 6 | **8** | 6 × 6 | **9** | 12 × 1 | **10** | 12 × 2 | **11** | 12 × 3 | **12** | 9 × 5 |

| **13** | 12 × 5 | **14** | 6 ×12 | **15** | 6 × 3 | **16** | 12 ×12 | **17** | 9 × 7 | **18** | 6 × 7 | **19** | 9 ×11 | **20** | 6 × 2 |

| **21** | 6 × 9 | **22** | 9 ×12 | **23** | 6 × 5 | **24** | 9 ×10 | **25** | 12 × 0 | **26** | 6 × 1 | **27** | 6 × 8 | **28** | 6 × 0 |

| **29** | 6 ×11 | **30** | 9 × 4 | **31** | 12 × 8 | **32** | 12 × 7 | **33** | 9 × 9 | **34** | 9 × 2 | **35** | 12 ×11 | **36** | 6 × 4 |

| **37** | 6 ×10 | **38** | 9 × 1 | **39** | 12 × 9 | **40** | 0 × 6 | **41** | 4 ×12 | **42** | 0 ×12 | **43** | 2 × 6 | **44** | 2 × 9 |

| **45** | 7 × 9 | **46** | 8 × 9 | **47** | 1 ×12 | **48** | 11 × 6 | **49** | 5 × 6 | **50** | 1 × 9 | **51** | 10 × 6 | **52** | 1 × 6 |

| **53** | 9 ×12 | **54** | 4 × 6 | **55** | 11 × 9 | **56** | 11 ×12 | **57** | 12 × 9 | **58** | 12 × 6 | **59** | 10 ×12 | **60** | 5 ×12 |

© 2024 Teach Elementals, LLC

MIXED REVIEW: 6, 9, AND 12, SET C

Name: _____

Date: _____

SCORE

1
$$\begin{array}{r} 6 \\ \times\ 3 \\ \hline \end{array}$$

2
$$\begin{array}{r} 6 \\ \times\ 1 \\ \hline \end{array}$$

3
$$\begin{array}{r} 9 \\ \times 10 \\ \hline \end{array}$$

4
$$\begin{array}{r} 12 \\ \times\ 9 \\ \hline \end{array}$$

5
$$\begin{array}{r} 9 \\ \times\ 7 \\ \hline \end{array}$$

6
$$\begin{array}{r} 9 \\ \times\ 8 \\ \hline \end{array}$$

7
$$\begin{array}{r} 6 \\ \times\ 8 \\ \hline \end{array}$$

8
$$\begin{array}{r} 12 \\ \times\ 5 \\ \hline \end{array}$$

9
$$\begin{array}{r} 6 \\ \times 10 \\ \hline \end{array}$$

10
$$\begin{array}{r} 12 \\ \times 12 \\ \hline \end{array}$$

11
$$\begin{array}{r} 9 \\ \times\ 4 \\ \hline \end{array}$$

12
$$\begin{array}{r} 9 \\ \times\ 9 \\ \hline \end{array}$$

13
$$\begin{array}{r} 12 \\ \times\ 8 \\ \hline \end{array}$$

14
$$\begin{array}{r} 9 \\ \times\ 0 \\ \hline \end{array}$$

15
$$\begin{array}{r} 6 \\ \times 12 \\ \hline \end{array}$$

16
$$\begin{array}{r} 12 \\ \times 11 \\ \hline \end{array}$$

17
$$\begin{array}{r} 9 \\ \times 12 \\ \hline \end{array}$$

18
$$\begin{array}{r} 9 \\ \times\ 2 \\ \hline \end{array}$$

19
$$\begin{array}{r} 6 \\ \times\ 2 \\ \hline \end{array}$$

20
$$\begin{array}{r} 12 \\ \times\ 7 \\ \hline \end{array}$$

21
$$\begin{array}{r} 12 \\ \times\ 6 \\ \hline \end{array}$$

22
$$\begin{array}{r} 9 \\ \times\ 1 \\ \hline \end{array}$$

23
$$\begin{array}{r} 12 \\ \times 10 \\ \hline \end{array}$$

24
$$\begin{array}{r} 6 \\ \times 11 \\ \hline \end{array}$$

25
$$\begin{array}{r} 12 \\ \times\ 3 \\ \hline \end{array}$$

26
$$\begin{array}{r} 6 \\ \times\ 7 \\ \hline \end{array}$$

27
$$\begin{array}{r} 12 \\ \times\ 2 \\ \hline \end{array}$$

28
$$\begin{array}{r} 6 \\ \times\ 5 \\ \hline \end{array}$$

29
$$\begin{array}{r} 6 \\ \times\ 0 \\ \hline \end{array}$$

30
$$\begin{array}{r} 12 \\ \times\ 0 \\ \hline \end{array}$$

31
$$\begin{array}{r} 6 \\ \times\ 9 \\ \hline \end{array}$$

32
$$\begin{array}{r} 12 \\ \times\ 1 \\ \hline \end{array}$$

33
$$\begin{array}{r} 9 \\ \times 11 \\ \hline \end{array}$$

34
$$\begin{array}{r} 6 \\ \times\ 6 \\ \hline \end{array}$$

35
$$\begin{array}{r} 9 \\ \times\ 5 \\ \hline \end{array}$$

36
$$\begin{array}{r} 6 \\ \times\ 4 \\ \hline \end{array}$$

37
$$\begin{array}{r} 12 \\ \times\ 4 \\ \hline \end{array}$$

38
$$\begin{array}{r} 9 \\ \times\ 6 \\ \hline \end{array}$$

39
$$\begin{array}{r} 9 \\ \times\ 3 \\ \hline \end{array}$$

40
$$\begin{array}{r} 5 \\ \times\ 6 \\ \hline \end{array}$$

41
$$\begin{array}{r} 11 \\ \times 12 \\ \hline \end{array}$$

42
$$\begin{array}{r} 11 \\ \times\ 9 \\ \hline \end{array}$$

43
$$\begin{array}{r} 9 \\ \times\ 6 \\ \hline \end{array}$$

44
$$\begin{array}{r} 3 \\ \times 12 \\ \hline \end{array}$$

45
$$\begin{array}{r} 10 \\ \times 12 \\ \hline \end{array}$$

46
$$\begin{array}{r} 4 \\ \times\ 6 \\ \hline \end{array}$$

47
$$\begin{array}{r} 7 \\ \times 12 \\ \hline \end{array}$$

48
$$\begin{array}{r} 1 \\ \times\ 6 \\ \hline \end{array}$$

49
$$\begin{array}{r} 3 \\ \times\ 6 \\ \hline \end{array}$$

50
$$\begin{array}{r} 1 \\ \times\ 9 \\ \hline \end{array}$$

51
$$\begin{array}{r} 4 \\ \times\ 9 \\ \hline \end{array}$$

52
$$\begin{array}{r} 12 \\ \times\ 6 \\ \hline \end{array}$$

53
$$\begin{array}{r} 1 \\ \times 12 \\ \hline \end{array}$$

54
$$\begin{array}{r} 6 \\ \times 12 \\ \hline \end{array}$$

55
$$\begin{array}{r} 10 \\ \times\ 9 \\ \hline \end{array}$$

56
$$\begin{array}{r} 4 \\ \times 12 \\ \hline \end{array}$$

57
$$\begin{array}{r} 11 \\ \times\ 6 \\ \hline \end{array}$$

58
$$\begin{array}{r} 0 \\ \times 12 \\ \hline \end{array}$$

59
$$\begin{array}{r} 0 \\ \times\ 9 \\ \hline \end{array}$$

60
$$\begin{array}{r} 7 \\ \times\ 9 \\ \hline \end{array}$$

© 2024 Teach Elementals, LLC

Name: _____

Date: _____

SCORE

1 2 × 7	**2** 6 × 7	**3** 5 × 3	**4** 3 × 7
5 7 × 2	**6** 3 × 6	**7** 0 × 3	**8** 3 × 8

(Items 9–12 continue row)

9 4 × 3	**10** 1 × 3	**11** 7 × 5	**12** 3 × 9
13 12 × 7	**14** 5 × 7	**15** 3 × 2	**16** 3 × 1
17 3 × 0	**18** 10 × 7	**19** 7 × 3	**20** 3 × 5
21 3 × 4	**22** 11 × 3	**23** 9 × 3	**24** 7 × 6
25 7 × 9	**26** 3 ×12	**27** 7 ×10	**28** 3 × 3
29 7 ×11	**30** 8 × 7	**31** 9 × 7	**32** 7 × 3
33 6 × 3	**34** 7 × 8	**35** 2 × 3	**36** 12 × 3
37 3 × 7	**38** 3 ×10	**39** 7 × 7	**40** 7 × 0
41 8 × 3	**42** 7 × 1	**43** 4 × 7	**44** 1 × 7
45 3 ×11	**46** 0 × 7	**47** 10 × 3	**48** 11 × 7
49 7 × 4	**50** 7 ×12	**51** 10 × 3	**52** 8 × 7
53 5 × 7	**54** 3 × 7	**55** 7 × 8	**56** 7 × 3
57 3 × 2	**58** 3 × 4	**59** 7 × 2	**60** 3 × 8

© 2024 Teach Elementals, LLC

Name: _____

Date: _____

SCORE

1	**2**	**3**	**4**
7 × 4	3 × 0	7 ×11	3 × 6

5	**6**	**7**	**8**	**9**	**10**	**11**	**12**
2 × 3	7 × 8	3 × 9	7 × 2	3 ×11	7 × 3	5 × 3	4 × 7

13	**14**	**15**	**16**	**17**	**18**	**19**	**20**
7 ×12	3 × 5	9 × 7	10 × 3	11 × 3	12 × 3	7 × 6	1 × 7

21	**22**	**23**	**24**	**25**	**26**	**27**	**28**
7 × 1	2 × 7	7 × 9	4 × 3	3 ×10	3 × 3	3 × 4	0 × 7

29	**30**	**31**	**32**	**33**	**34**	**35**	**36**
8 × 3	3 × 2	7 ×10	3 × 7	7 × 5	1 × 3	7 × 7	10 × 7

37	**38**	**39**	**40**	**41**	**42**	**43**	**44**
3 ×12	7 × 3	6 × 7	6 × 3	11 × 7	0 × 3	7 × 0	3 × 7

45	**46**	**47**	**48**	**49**	**50**	**51**	**52**
5 × 7	8 × 7	3 × 8	9 × 3	12 × 7	3 × 1	7 × 6	0 × 7

53	**54**	**55**	**56**	**57**	**58**	**59**	**60**
7 × 9	3 ×12	3 × 3	7 × 2	4 × 3	7 × 5	10 × 7	2 × 3

© 2024 Teach Elementals, LLC

Name: _____

Date: _____

SCORE

1 $\begin{array}{r} 5 \\ \times\ 3 \\ \hline \end{array}$	**2** $\begin{array}{r} 7 \\ \times 10 \\ \hline \end{array}$	**3** $\begin{array}{r} 0 \\ \times\ 6 \\ \hline \end{array}$	**4** $\begin{array}{r} 1 \\ \times\ 8 \\ \hline \end{array}$

5 $\begin{array}{r} 1 \\ \times\ 1 \\ \hline \end{array}$	**6** $\begin{array}{r} 8 \\ \times 12 \\ \hline \end{array}$	**7** $\begin{array}{r} 7 \\ \times\ 3 \\ \hline \end{array}$	**8** $\begin{array}{r} 7 \\ \times 11 \\ \hline \end{array}$	**9** $\begin{array}{r} 9 \\ \times\ 6 \\ \hline \end{array}$	**10** $\begin{array}{r} 10 \\ \times\ 9 \\ \hline \end{array}$	**11** $\begin{array}{r} 11 \\ \times 11 \\ \hline \end{array}$
12 $\begin{array}{r} 5 \\ \times 11 \\ \hline \end{array}$						
13 $\begin{array}{r} 4 \\ \times\ 5 \\ \hline \end{array}$	**14** $\begin{array}{r} 6 \\ \times\ 1 \\ \hline \end{array}$	**15** $\begin{array}{r} 9 \\ \times\ 2 \\ \hline \end{array}$	**16** $\begin{array}{r} 10 \\ \times\ 5 \\ \hline \end{array}$	**17** $\begin{array}{r} 6 \\ \times\ 6 \\ \hline \end{array}$	**18** $\begin{array}{r} 6 \\ \times\ 7 \\ \hline \end{array}$	**19** $\begin{array}{r} 2 \\ \times\ 3 \\ \hline \end{array}$
20 $\begin{array}{r} 12 \\ \times\ 8 \\ \hline \end{array}$						
21 $\begin{array}{r} 3 \\ \times\ 5 \\ \hline \end{array}$	**22** $\begin{array}{r} 5 \\ \times 12 \\ \hline \end{array}$	**23** $\begin{array}{r} 9 \\ \times\ 7 \\ \hline \end{array}$	**24** $\begin{array}{r} 7 \\ \times\ 4 \\ \hline \end{array}$	**25** $\begin{array}{r} 0 \\ \times\ 0 \\ \hline \end{array}$	**26** $\begin{array}{r} 7 \\ \times\ 2 \\ \hline \end{array}$	**27** $\begin{array}{r} 5 \\ \times 10 \\ \hline \end{array}$
28 $\begin{array}{r} 9 \\ \times\ 9 \\ \hline \end{array}$						
29 $\begin{array}{r} 6 \\ \times\ 3 \\ \hline \end{array}$	**30** $\begin{array}{r} 6 \\ \times 12 \\ \hline \end{array}$	**31** $\begin{array}{r} 5 \\ \times\ 2 \\ \hline \end{array}$	**32** $\begin{array}{r} 11 \\ \times\ 8 \\ \hline \end{array}$	**33** $\begin{array}{r} 12 \\ \times\ 7 \\ \hline \end{array}$	**34** $\begin{array}{r} 9 \\ \times\ 4 \\ \hline \end{array}$	**35** $\begin{array}{r} 9 \\ \times 11 \\ \hline \end{array}$
36 $\begin{array}{r} 12 \\ \times\ 6 \\ \hline \end{array}$						
37 $\begin{array}{r} 12 \\ \times 10 \\ \hline \end{array}$	**38** $\begin{array}{r} 3 \\ \times\ 9 \\ \hline \end{array}$	**39** $\begin{array}{r} 11 \\ \times 12 \\ \hline \end{array}$	**40** $\begin{array}{r} 2 \\ \times\ 9 \\ \hline \end{array}$	**41** $\begin{array}{r} 12 \\ \times\ 9 \\ \hline \end{array}$	**42** $\begin{array}{r} 12 \\ \times\ 2 \\ \hline \end{array}$	**43** $\begin{array}{r} 7 \\ \times\ 6 \\ \hline \end{array}$
44 $\begin{array}{r} 12 \\ \times 11 \\ \hline \end{array}$						
45 $\begin{array}{r} 4 \\ \times\ 6 \\ \hline \end{array}$	**46** $\begin{array}{r} 4 \\ \times\ 8 \\ \hline \end{array}$	**47** $\begin{array}{r} 5 \\ \times\ 7 \\ \hline \end{array}$	**48** $\begin{array}{r} 0 \\ \times\ 9 \\ \hline \end{array}$	**49** $\begin{array}{r} 7 \\ \times\ 9 \\ \hline \end{array}$	**50** $\begin{array}{r} 5 \\ \times\ 4 \\ \hline \end{array}$	**51** $\begin{array}{r} 11 \\ \times\ 4 \\ \hline \end{array}$
52 $\begin{array}{r} 3 \\ \times\ 6 \\ \hline \end{array}$						
53 $\begin{array}{r} 8 \\ \times\ 2 \\ \hline \end{array}$	**54** $\begin{array}{r} 11 \\ \times\ 2 \\ \hline \end{array}$	**55** $\begin{array}{r} 5 \\ \times\ 9 \\ \hline \end{array}$	**56** $\begin{array}{r} 3 \\ \times\ 4 \\ \hline \end{array}$	**57** $\begin{array}{r} 11 \\ \times\ 9 \\ \hline \end{array}$	**58** $\begin{array}{r} 3 \\ \times\ 8 \\ \hline \end{array}$	**59** $\begin{array}{r} 7 \\ \times\ 7 \\ \hline \end{array}$
60 $\begin{array}{r} 12 \\ \times\ 5 \\ \hline \end{array}$						

© 2024 Teach Elementals, LLC

Name: _____

Date: _____

SCORE

1
5
× 8

2
2
× 2

3
7
× 4

4
8
× 4

5
3
× 6

6
10
× 7

7
2
× 1

8
6
× 3

9
6
× 9

10
5
× 6

11
9
× 3

12
3
×12

13
6
× 2

14
7
× 2

15
6
×11

16
9
×12

17
4
×11

18
8
× 0

19
3
× 3

20
6
× 5

21
7
× 8

22
3
× 4

23
9
× 7

24
8
×10

25
11
× 9

26
10
× 2

27
6
×10

28
8
×11

29
3
×10

30
10
× 8

31
2
× 4

32
6
×12

33
10
×11

34
7
×12

35
2
× 7

36
0
×11

37
9
× 2

38
10
×10

39
5
× 2

40
12
× 4

41
3
× 8

42
11
×10

43
11
× 4

44
6
× 4

45
6
× 8

46
5
×12

47
0
× 1

48
12
×12

49
7
× 6

50
10
×12

51
4
× 4

52
0
× 2

53
11
× 5

54
3
×11

55
10
× 3

56
9
×11

57
1
× 3

58
11
×11

59
12
× 3

60
8
× 6

© 2024 Teach Elementals, LLC

Name: _____   Date: _____

SCORE

1
6
×11

2
12
×10

3
5
×11

4
7
× 9

5
12
× 3

6
10
× 9

7
12
×12

8
10
×12

9
8
×10

10
10
× 2

11
3
× 3

12
4
× 0

13
3
× 2

14
7
× 5

15
10
× 5

16
7
× 8

17
8
×12

18
9
×10

19
2
×10

20
11
× 2

21
4
× 9

22
9
× 8

23
0
× 5

24
10
× 4

25
4
× 1

26
1
× 7

27
5
× 8

28
6
×10

29
5
× 9

30
8
× 3

31
12
× 8

32
11
× 1

33
3
× 7

34
11
× 6

35
10
×11

36
3
×10

37
3
×12

38
11
× 7

39
11
× 3

40
8
× 9

41
8
× 4

42
4
× 3

43
9
× 5

44
5
× 3

45
6
× 9

46
8
× 7

47
11
× 8

48
6
× 6

49
2
×12

50
0
×12

51
8
× 8

52
11
×12

53
12
×11

54
5
× 5

55
8
×11

56
2
× 5

57
9
× 4

58
10
× 6

59
3
× 5

60
8
× 5

© 2024 Teach Elementals, LLC

MIXED REVIEW: ALL FACTS, SET D

Name: _____   Date: _____

SCORE

1 $\begin{array}{r} 9 \\ \times\ 1 \\ \hline \end{array}$	**2** $\begin{array}{r} 0 \\ \times\ 1 \\ \hline \end{array}$	**3** $\begin{array}{r} 10 \\ \times\ 2 \\ \hline \end{array}$	**4** $\begin{array}{r} 12 \\ \times\ 1 \\ \hline \end{array}$

5 $\begin{array}{r} 3 \\ \times\ 5 \\ \hline \end{array}$	**6** $\begin{array}{r} 4 \\ \times 10 \\ \hline \end{array}$	**7** $\begin{array}{r} 7 \\ \times\ 2 \\ \hline \end{array}$	**8** $\begin{array}{r} 12 \\ \times\ 8 \\ \hline \end{array}$	**9** $\begin{array}{r} 8 \\ \times\ 2 \\ \hline \end{array}$	**10** $\begin{array}{r} 1 \\ \times 10 \\ \hline \end{array}$	**11** $\begin{array}{r} 4 \\ \times\ 7 \\ \hline \end{array}$	**12** $\begin{array}{r} 6 \\ \times 12 \\ \hline \end{array}$
13 $\begin{array}{r} 8 \\ \times\ 7 \\ \hline \end{array}$	**14** $\begin{array}{r} 12 \\ \times\ 6 \\ \hline \end{array}$	**15** $\begin{array}{r} 7 \\ \times 10 \\ \hline \end{array}$	**16** $\begin{array}{r} 12 \\ \times\ 3 \\ \hline \end{array}$	**17** $\begin{array}{r} 11 \\ \times\ 2 \\ \hline \end{array}$	**18** $\begin{array}{r} 10 \\ \times\ 3 \\ \hline \end{array}$	**19** $\begin{array}{r} 8 \\ \times\ 4 \\ \hline \end{array}$	**20** $\begin{array}{r} 11 \\ \times\ 5 \\ \hline \end{array}$
21 $\begin{array}{r} 4 \\ \times\ 2 \\ \hline \end{array}$	**22** $\begin{array}{r} 5 \\ \times 11 \\ \hline \end{array}$	**23** $\begin{array}{r} 3 \\ \times\ 4 \\ \hline \end{array}$	**24** $\begin{array}{r} 7 \\ \times\ 0 \\ \hline \end{array}$	**25** $\begin{array}{r} 3 \\ \times 10 \\ \hline \end{array}$	**26** $\begin{array}{r} 12 \\ \times\ 5 \\ \hline \end{array}$	**27** $\begin{array}{r} 9 \\ \times\ 6 \\ \hline \end{array}$	**28** $\begin{array}{r} 7 \\ \times\ 5 \\ \hline \end{array}$
29 $\begin{array}{r} 0 \\ \times 10 \\ \hline \end{array}$	**30** $\begin{array}{r} 9 \\ \times 11 \\ \hline \end{array}$	**31** $\begin{array}{r} 7 \\ \times\ 8 \\ \hline \end{array}$	**32** $\begin{array}{r} 9 \\ \times\ 4 \\ \hline \end{array}$	**33** $\begin{array}{r} 3 \\ \times\ 2 \\ \hline \end{array}$	**34** $\begin{array}{r} 3 \\ \times\ 3 \\ \hline \end{array}$	**35** $\begin{array}{r} 6 \\ \times\ 4 \\ \hline \end{array}$	**36** $\begin{array}{r} 7 \\ \times\ 7 \\ \hline \end{array}$
37 $\begin{array}{r} 3 \\ \times 11 \\ \hline \end{array}$	**38** $\begin{array}{r} 8 \\ \times\ 5 \\ \hline \end{array}$	**39** $\begin{array}{r} 9 \\ \times\ 9 \\ \hline \end{array}$	**40** $\begin{array}{r} 5 \\ \times\ 2 \\ \hline \end{array}$	**41** $\begin{array}{r} 12 \\ \times 10 \\ \hline \end{array}$	**42** $\begin{array}{r} 10 \\ \times\ 6 \\ \hline \end{array}$	**43** $\begin{array}{r} 12 \\ \times\ 7 \\ \hline \end{array}$	**44** $\begin{array}{r} 5 \\ \times\ 5 \\ \hline \end{array}$
45 $\begin{array}{r} 6 \\ \times\ 6 \\ \hline \end{array}$	**46** $\begin{array}{r} 10 \\ \times 10 \\ \hline \end{array}$	**47** $\begin{array}{r} 10 \\ \times 12 \\ \hline \end{array}$	**48** $\begin{array}{r} 12 \\ \times 12 \\ \hline \end{array}$	**49** $\begin{array}{r} 11 \\ \times\ 4 \\ \hline \end{array}$	**50** $\begin{array}{r} 2 \\ \times\ 6 \\ \hline \end{array}$	**51** $\begin{array}{r} 9 \\ \times 12 \\ \hline \end{array}$	**52** $\begin{array}{r} 6 \\ \times 10 \\ \hline \end{array}$
53 $\begin{array}{r} 2 \\ \times\ 8 \\ \hline \end{array}$	**54** $\begin{array}{r} 11 \\ \times\ 7 \\ \hline \end{array}$	**55** $\begin{array}{r} 8 \\ \times\ 3 \\ \hline \end{array}$	**56** $\begin{array}{r} 6 \\ \times\ 8 \\ \hline \end{array}$	**57** $\begin{array}{r} 12 \\ \times\ 2 \\ \hline \end{array}$	**58** $\begin{array}{r} 9 \\ \times\ 7 \\ \hline \end{array}$	**59** $\begin{array}{r} 8 \\ \times 12 \\ \hline \end{array}$	**60** $\begin{array}{r} 9 \\ \times\ 5 \\ \hline \end{array}$

© 2024 Teach Elementals, LLC

Name: _____ Date: _____

SCORE

1 5 ×12	**2** 2 × 6	**3** 3 × 6	**4** 6 × 4

5 5 × 3	**6** 2 × 4	**7** 0 × 0	**8** 11 ×10	**9** 12 × 9	**10** 10 ×10	**11** 0 ×12	**12** 8 × 4

| **13** 9 × 9 | **14** 1 × 9 | **15** 3 ×12 | **16** 5 ×10 | **17** 7 × 6 | **18** 9 ×12 | **19** 5 × 2 | **20** 11 ×12 |

| **21** 7 × 3 | **22** 5 × 1 | **23** 2 ×10 | **24** 3 × 5 | **25** 9 × 2 | **26** 3 ×10 | **27** 8 ×11 | **28** 10 × 5 |

| **29** 5 × 7 | **30** 10 × 4 | **31** 12 × 1 | **32** 10 × 8 | **33** 11 × 9 | **34** 7 × 5 | **35** 11 × 8 | **36** 6 ×12 |

| **37** 12 ×10 | **38** 5 × 5 | **39** 6 × 8 | **40** 9 × 3 | **41** 4 × 2 | **42** 3 × 7 | **43** 7 ×12 | **44** 5 ×11 |

| **45** 11 ×11 | **46** 8 × 6 | **47** 6 × 5 | **48** 2 × 9 | **49** 10 ×11 | **50** 8 ×12 | **51** 7 × 9 | **52** 12 ×11 |

| **53** 7 × 7 | **54** 8 ×10 | **55** 5 × 6 | **56** 9 ×10 | **57** 3 × 8 | **58** 10 × 9 | **59** 0 ×10 | **60** 6 × 7 |

© 2024 Teach Elementals, LLC

Name: _____

Date: _____

SCORE

1 $\begin{array}{r} 5 \\ \times\ 9 \\ \hline \end{array}$	**2** $\begin{array}{r} 3 \\ \times\ 3 \\ \hline \end{array}$	**3** $\begin{array}{r} 5 \\ \times 10 \\ \hline \end{array}$	**4** $\begin{array}{r} 11 \\ \times 11 \\ \hline \end{array}$

5 $\begin{array}{r} 9 \\ \times 10 \\ \hline \end{array}$	**6** $\begin{array}{r} 11 \\ \times\ 7 \\ \hline \end{array}$	**7** $\begin{array}{r} 11 \\ \times\ 2 \\ \hline \end{array}$	**8** $\begin{array}{r} 12 \\ \times\ 4 \\ \hline \end{array}$	**9** $\begin{array}{r} 7 \\ \times\ 8 \\ \hline \end{array}$	**10** $\begin{array}{r} 1 \\ \times\ 4 \\ \hline \end{array}$	**11** $\begin{array}{r} 6 \\ \times\ 0 \\ \hline \end{array}$	**12** $\begin{array}{r} 5 \\ \times\ 7 \\ \hline \end{array}$
13 $\begin{array}{r} 9 \\ \times\ 8 \\ \hline \end{array}$	**14** $\begin{array}{r} 6 \\ \times 10 \\ \hline \end{array}$	**15** $\begin{array}{r} 6 \\ \times 11 \\ \hline \end{array}$	**16** $\begin{array}{r} 11 \\ \times\ 3 \\ \hline \end{array}$	**17** $\begin{array}{r} 6 \\ \times\ 9 \\ \hline \end{array}$	**18** $\begin{array}{r} 0 \\ \times\ 1 \\ \hline \end{array}$	**19** $\begin{array}{r} 3 \\ \times\ 9 \\ \hline \end{array}$	**20** $\begin{array}{r} 12 \\ \times\ 3 \\ \hline \end{array}$
21 $\begin{array}{r} 6 \\ \times\ 5 \\ \hline \end{array}$	**22** $\begin{array}{r} 3 \\ \times\ 6 \\ \hline \end{array}$	**23** $\begin{array}{r} 2 \\ \times 11 \\ \hline \end{array}$	**24** $\begin{array}{r} 10 \\ \times\ 6 \\ \hline \end{array}$	**25** $\begin{array}{r} 5 \\ \times 12 \\ \hline \end{array}$	**26** $\begin{array}{r} 5 \\ \times\ 8 \\ \hline \end{array}$	**27** $\begin{array}{r} 11 \\ \times\ 6 \\ \hline \end{array}$	**28** $\begin{array}{r} 12 \\ \times 12 \\ \hline \end{array}$
29 $\begin{array}{r} 6 \\ \times\ 7 \\ \hline \end{array}$	**30** $\begin{array}{r} 1 \\ \times\ 1 \\ \hline \end{array}$	**31** $\begin{array}{r} 11 \\ \times\ 8 \\ \hline \end{array}$	**32** $\begin{array}{r} 10 \\ \times\ 7 \\ \hline \end{array}$	**33** $\begin{array}{r} 10 \\ \times\ 8 \\ \hline \end{array}$	**34** $\begin{array}{r} 4 \\ \times 11 \\ \hline \end{array}$	**35** $\begin{array}{r} 4 \\ \times\ 5 \\ \hline \end{array}$	**36** $\begin{array}{r} 9 \\ \times\ 2 \\ \hline \end{array}$
37 $\begin{array}{r} 9 \\ \times\ 6 \\ \hline \end{array}$	**38** $\begin{array}{r} 7 \\ \times 11 \\ \hline \end{array}$	**39** $\begin{array}{r} 8 \\ \times\ 7 \\ \hline \end{array}$	**40** $\begin{array}{r} 11 \\ \times 10 \\ \hline \end{array}$	**41** $\begin{array}{r} 10 \\ \times\ 4 \\ \hline \end{array}$	**42** $\begin{array}{r} 6 \\ \times\ 3 \\ \hline \end{array}$	**43** $\begin{array}{r} 6 \\ \times\ 2 \\ \hline \end{array}$	**44** $\begin{array}{r} 10 \\ \times\ 2 \\ \hline \end{array}$
45 $\begin{array}{r} 4 \\ \times 10 \\ \hline \end{array}$	**46** $\begin{array}{r} 12 \\ \times\ 7 \\ \hline \end{array}$	**47** $\begin{array}{r} 8 \\ \times\ 5 \\ \hline \end{array}$	**48** $\begin{array}{r} 10 \\ \times\ 9 \\ \hline \end{array}$	**49** $\begin{array}{r} 9 \\ \times\ 3 \\ \hline \end{array}$	**50** $\begin{array}{r} 3 \\ \times 11 \\ \hline \end{array}$	**51** $\begin{array}{r} 5 \\ \times\ 4 \\ \hline \end{array}$	**52** $\begin{array}{r} 7 \\ \times\ 4 \\ \hline \end{array}$
53 $\begin{array}{r} 8 \\ \times\ 9 \\ \hline \end{array}$	**54** $\begin{array}{r} 0 \\ \times\ 3 \\ \hline \end{array}$	**55** $\begin{array}{r} 9 \\ \times\ 7 \\ \hline \end{array}$	**56** $\begin{array}{r} 0 \\ \times\ 7 \\ \hline \end{array}$	**57** $\begin{array}{r} 8 \\ \times\ 8 \\ \hline \end{array}$	**58** $\begin{array}{r} 8 \\ \times 11 \\ \hline \end{array}$	**59** $\begin{array}{r} 11 \\ \times\ 4 \\ \hline \end{array}$	**60** $\begin{array}{r} 4 \\ \times\ 8 \\ \hline \end{array}$

© 2024 Teach Elementals, LLC

Name: _____   Date: _____

SCORE

1 10 × 6	**2** 3 × 2	**3** 5 × 5	**4** 11 ×10
5 2 × 3	**6** 8 × 6	**7** 0 × 2	**8** 9 × 4
9 11 × 5	**10** 8 × 3	**11** 12 ×10	**12** 8 × 7
13 8 ×10	**14** 8 ×12	**15** 10 × 2	**16** 10 × 9
17 9 × 2	**18** 3 ×10	**19** 1 × 8	**20** 9 × 9
21 7 × 5	**22** 6 × 9	**23** 7 ×10	**24** 2 × 2
25 3 × 8	**26** 5 × 7	**27** 11 ×11	**28** 3 × 7
29 9 × 3	**30** 9 × 6	**31** 11 × 7	**32** 6 × 2
33 9 × 7	**34** 10 × 1	**35** 9 × 0	**36** 6 ×11
37 6 × 7	**38** 3 × 5	**39** 11 × 3	**40** 6 × 5
41 5 ×11	**42** 4 ×12	**43** 8 × 2	**44** 3 × 6
45 0 ×11	**46** 3 ×12	**47** 7 × 1	**48** 6 × 8
49 8 × 4	**50** 12 × 7	**51** 11 ×12	**52** 7 × 8
53 4 × 3	**54** 9 ×10	**55** 5 × 2	**56** 12 × 3
57 9 × 8	**58** 11 × 6	**59** 7 ×12	**60** 4 × 6

© 2024 Teach Elementals, LLC

Name: _____   Date: _____

SCORE

1 10 × 9	**2** 0 × 1	**3** 5 ×11	**4** 11 × 4
5 3 × 7	**6** 0 × 8	**7** 7 × 6	**8** 12 ×12
9 4 × 9	**10** 10 × 8	**11** 12 × 2	**12** 6 × 1
13 2 × 7	**14** 5 × 6	**15** 11 × 7	**16** 8 × 5
17 9 × 5	**18** 7 × 9	**19** 3 ×10	**20** 8 ×11
21 11 × 9	**22** 3 × 3	**23** 3 × 9	**24** 11 × 5
25 10 ×12	**26** 3 × 2	**27** 6 ×10	**28** 11 × 2
29 1 × 2	**30** 7 × 3	**31** 12 × 9	**32** 6 ×11
33 8 × 7	**34** 10 × 6	**35** 9 ×10	**36** 9 × 3
37 5 ×12	**38** 3 × 4	**39** 10 × 2	**40** 8 × 8
41 2 × 8	**42** 7 ×10	**43** 3 ×11	**44** 4 × 7
45 4 × 0	**46** 6 × 3	**47** 7 × 2	**48** 10 × 4
49 3 × 1	**50** 12 × 5	**51** 6 × 2	**52** 6 × 6
53 2 ×12	**54** 12 ×11	**55** 5 × 9	**56** 3 × 6
57 10 ×11	**58** 5 × 3	**59** 12 × 8	**60** 5 × 8

© 2024 Teach Elementals, LLC

Name: _____   Date: _____

SCORE

#		#		#		#	
1	0 ×3	**2**	9 ×9	**3**	7 ×7	**4**	10 ×8
5	10 ×10	**6**	10 ×12	**7**	11 ×2	**8**	11 ×10
9	8 ×2	**10**	9 ×11	**11**	11 ×8	**12**	2 ×12
13	5 ×10	**14**	5 ×7	**15**	10 ×11	**16**	5 ×6
17	9 ×4	**18**	12 ×6	**19**	3 ×5	**20**	8 ×9
21	12 ×11	**22**	1 ×8	**23**	12 ×8	**24**	10 ×7
25	9 ×8	**26**	7 ×4	**27**	5 ×4	**28**	12 ×9
29	9 ×12	**30**	9 ×5	**31**	11 ×11	**32**	4 ×5
33	11 ×9	**34**	8 ×12	**35**	6 ×5	**36**	8 ×3
37	4 ×11	**38**	7 ×9	**39**	12 ×2	**40**	4 ×2
41	2 ×1	**42**	11 ×3	**43**	10 ×5	**44**	7 ×2
45	7 ×8	**46**	9 ×7	**47**	6 ×4	**48**	5 ×12
49	6 ×3	**50**	10 ×3	**51**	9 ×0	**52**	3 ×8
53	3 ×12	**54**	6 ×12	**55**	12 ×4	**56**	2 ×9
57	0 ×2	**58**	7 ×11	**59**	3 ×3	**60**	7 ×1

© 2024 Teach Elementals, LLC

MIXED REVIEW: ALL FACTS, SET J

Name: _____

Date: _____

SCORE

1	2	3	4
4 × 1	10 × 7	5 × 5	8 × 3

5	6	7	8	9	10	11	12
7 × 3	3 ×12	2 × 2	4 ×12	5 ×12	12 ×11	3 × 8	12 ×12

13	14	15	16	17	18	19	20
8 × 4	0 × 7	10 × 5	2 × 6	3 × 3	12 × 5	9 × 4	8 × 7

21	22	23	24	25	26	27	28
12 × 4	3 × 6	9 × 6	4 ×10	5 × 8	11 × 5	9 × 5	5 ×11

29	30	31	32	33	34	35	36
7 ×11	2 × 4	10 × 9	8 ×11	12 ×10	12 × 0	11 × 8	10 × 1

37	38	39	40	41	42	43	44
3 × 4	9 ×11	8 × 8	6 × 2	1 ×11	7 × 9	11 × 4	10 × 6

45	46	47	48	49	50	51	52
5 × 3	7 ×10	0 × 4	6 × 9	9 ×12	9 × 7	7 × 2	11 × 7

53	54	55	56	57	58	59	60
10 × 3	11 × 2	12 × 3	12 × 7	5 × 6	5 ×10	7 × 4	11 × 3

© 2024 Teach Elementals, LLC

Name: _____ Date: _____

SCORE

1
3
× 4

2
9
× 8

3
8
× 5

4
10
× 4

5
10
×11

6
5
× 4

7
2
× 7

8
3
×10

9
12
× 1

10
10
× 8

11
5
× 2

12
3
× 3

13
8
× 4

14
6
× 3

15
8
×12

16
0
× 6

17
12
× 3

18
12
× 7

19
10
×10

20
3
× 5

21
2
× 8

22
8
× 6

23
12
×11

24
1
× 0

25
5
× 1

26
6
× 5

27
10
× 5

28
6
× 4

29
4
× 8

30
12
× 5

31
11
×10

32
10
× 6

33
0
× 0

34
7
×12

35
7
× 5

36
6
× 8

37
6
×10

38
12
×12

39
3
× 2

40
0
×11

41
11
× 9

42
12
× 8

43
11
×12

44
12
× 6

45
6
× 7

46
6
× 6

47
8
×10

48
7
× 6

49
10
× 3

50
7
×11

51
3
× 8

52
10
× 7

53
7
× 9

54
10
×12

55
2
×10

56
11
× 7

57
4
× 9

58
9
× 9

59
9
× 3

60
12
× 2

© 2024 Teach Elementals, LLC

MIXED REVIEW: ALL FACTS, SET L

Name: _____   Date: _____

SCORE

1 $\begin{array}{r} 6 \\ \times\ 1 \\ \hline \end{array}$	**2** $\begin{array}{r} 3 \\ \times\ 9 \\ \hline \end{array}$	**3** $\begin{array}{r} 10 \\ \times\ 8 \\ \hline \end{array}$	**4** $\begin{array}{r} 7 \\ \times 10 \\ \hline \end{array}$

5 $\begin{array}{r} 8 \\ \times\ 9 \\ \hline \end{array}$	**6** $\begin{array}{r} 4 \\ \times\ 4 \\ \hline \end{array}$	**7** $\begin{array}{r} 5 \\ \times\ 6 \\ \hline \end{array}$	**8** $\begin{array}{r} 7 \\ \times\ 8 \\ \hline \end{array}$	**9** $\begin{array}{r} 11 \\ \times\ 2 \\ \hline \end{array}$	**10** $\begin{array}{r} 9 \\ \times\ 3 \\ \hline \end{array}$	**11** $\begin{array}{r} 5 \\ \times\ 9 \\ \hline \end{array}$	**12** $\begin{array}{r} 12 \\ \times\ 4 \\ \hline \end{array}$
13 $\begin{array}{r} 6 \\ \times 11 \\ \hline \end{array}$	**14** $\begin{array}{r} 7 \\ \times\ 6 \\ \hline \end{array}$	**15** $\begin{array}{r} 5 \\ \times 10 \\ \hline \end{array}$	**16** $\begin{array}{r} 1 \\ \times\ 1 \\ \hline \end{array}$	**17** $\begin{array}{r} 9 \\ \times\ 9 \\ \hline \end{array}$	**18** $\begin{array}{r} 6 \\ \times 12 \\ \hline \end{array}$	**19** $\begin{array}{r} 2 \\ \times 11 \\ \hline \end{array}$	**20** $\begin{array}{r} 4 \\ \times\ 7 \\ \hline \end{array}$
21 $\begin{array}{r} 8 \\ \times\ 3 \\ \hline \end{array}$	**22** $\begin{array}{r} 7 \\ \times\ 2 \\ \hline \end{array}$	**23** $\begin{array}{r} 6 \\ \times\ 7 \\ \hline \end{array}$	**24** $\begin{array}{r} 0 \\ \times\ 8 \\ \hline \end{array}$	**25** $\begin{array}{r} 5 \\ \times 12 \\ \hline \end{array}$	**26** $\begin{array}{r} 3 \\ \times\ 7 \\ \hline \end{array}$	**27** $\begin{array}{r} 11 \\ \times\ 6 \\ \hline \end{array}$	**28** $\begin{array}{r} 0 \\ \times\ 5 \\ \hline \end{array}$
29 $\begin{array}{r} 5 \\ \times\ 7 \\ \hline \end{array}$	**30** $\begin{array}{r} 9 \\ \times\ 4 \\ \hline \end{array}$	**31** $\begin{array}{r} 3 \\ \times\ 5 \\ \hline \end{array}$	**32** $\begin{array}{r} 4 \\ \times\ 6 \\ \hline \end{array}$	**33** $\begin{array}{r} 12 \\ \times\ 9 \\ \hline \end{array}$	**34** $\begin{array}{r} 9 \\ \times 11 \\ \hline \end{array}$	**35** $\begin{array}{r} 11 \\ \times 11 \\ \hline \end{array}$	**36** $\begin{array}{r} 9 \\ \times\ 1 \\ \hline \end{array}$
37 $\begin{array}{r} 12 \\ \times 10 \\ \hline \end{array}$	**38** $\begin{array}{r} 8 \\ \times\ 2 \\ \hline \end{array}$	**39** $\begin{array}{r} 9 \\ \times\ 8 \\ \hline \end{array}$	**40** $\begin{array}{r} 8 \\ \times\ 6 \\ \hline \end{array}$	**41** $\begin{array}{r} 11 \\ \times\ 9 \\ \hline \end{array}$	**42** $\begin{array}{r} 6 \\ \times\ 8 \\ \hline \end{array}$	**43** $\begin{array}{r} 7 \\ \times\ 5 \\ \hline \end{array}$	**44** $\begin{array}{r} 3 \\ \times\ 2 \\ \hline \end{array}$
45 $\begin{array}{r} 11 \\ \times 10 \\ \hline \end{array}$	**46** $\begin{array}{r} 10 \\ \times\ 2 \\ \hline \end{array}$	**47** $\begin{array}{r} 3 \\ \times 11 \\ \hline \end{array}$	**48** $\begin{array}{r} 8 \\ \times\ 8 \\ \hline \end{array}$	**49** $\begin{array}{r} 8 \\ \times 12 \\ \hline \end{array}$	**50** $\begin{array}{r} 6 \\ \times\ 9 \\ \hline \end{array}$	**51** $\begin{array}{r} 9 \\ \times 10 \\ \hline \end{array}$	**52** $\begin{array}{r} 8 \\ \times\ 5 \\ \hline \end{array}$
53 $\begin{array}{r} 8 \\ \times 11 \\ \hline \end{array}$	**54** $\begin{array}{r} 6 \\ \times\ 5 \\ \hline \end{array}$	**55** $\begin{array}{r} 7 \\ \times\ 7 \\ \hline \end{array}$	**56** $\begin{array}{r} 5 \\ \times\ 3 \\ \hline \end{array}$	**57** $\begin{array}{r} 9 \\ \times\ 2 \\ \hline \end{array}$	**58** $\begin{array}{r} 2 \\ \times\ 3 \\ \hline \end{array}$	**59** $\begin{array}{r} 8 \\ \times\ 7 \\ \hline \end{array}$	**60** $\begin{array}{r} 10 \\ \times\ 0 \\ \hline \end{array}$

© 2024 Teach Elementals, LLC

Name: _____   Date: _____

SCORE

1
12
× 4

2
3
× 8

3
5
× 8

4
12
×11

5
5
×11

6
0
× 6

7
11
× 6

8
12
× 2

9
6
× 9

10
6
×12

11
5
× 9

12
3
×12

13
12
× 6

14
5
×12

15
1
× 5

16
7
× 8

17
3
× 7

18
2
× 7

19
5
× 4

20
9
× 4

21
7
×10

22
12
× 9

23
3
× 1

24
4
×10

25
0
× 7

26
10
×12

27
12
× 3

28
2
× 4

29
11
×12

30
10
× 5

31
8
× 9

32
3
× 4

33
3
×11

34
8
×12

35
4
× 5

36
11
× 5

37
6
× 7

38
6
× 3

39
2
×12

40
5
× 6

41
7
× 4

42
9
×11

43
3
× 6

44
6
× 5

45
9
× 7

46
8
× 8

47
12
× 1

48
12
×12

49
8
× 4

50
11
× 2

51
12
× 7

52
7
× 2

53
6
× 2

54
8
× 2

55
8
×11

56
3
× 0

57
6
×10

58
9
× 8

59
6
× 8

60
9
× 6

Name: _____   Date: _____

SCORE

1 11 × 9	**2** 7 × 2	**3** 7 × 9	**4** 12 ×10
5 4 × 6	**6** 3 × 7	**7** 8 × 9	**8** 10 ×10
9 10 × 6	**10** 9 × 2	**11** 8 × 4	**12** 6 × 2
13 5 × 5	**14** 10 × 2	**15** 11 × 0	**16** 4 × 9
17 7 × 6	**18** 7 × 3	**19** 12 ×11	**20** 8 ×10
21 5 × 7	**22** 7 ×12	**23** 11 × 3	**24** 12 × 6
25 5 ×11	**26** 12 × 3	**27** 9 × 3	**28** 3 × 9
29 11 ×11	**30** 6 × 1	**31** 9 × 9	**32** 10 × 9
33 0 × 4	**34** 2 × 6	**35** 2 × 9	**36** 10 × 3
37 11 × 6	**38** 4 × 4	**39** 6 ×11	**40** 10 × 4
41 12 × 8	**42** 3 ×12	**43** 9 × 6	**44** 8 × 3
45 7 × 7	**46** 0 × 9	**47** 11 × 7	**48** 6 ×12
49 3 × 6	**50** 8 × 5	**51** 3 × 3	**52** 5 ×10
53 3 × 4	**54** 1 × 1	**55** 11 × 8	**56** 11 × 4
57 1 × 9	**58** 8 × 8	**59** 7 × 5	**60** 9 ×12

© 2024 Teach Elementals, LLC

Name: _____ Date: _____

SCORE

1
$\begin{array}{r} 9 \\ \times 11 \\ \hline \end{array}$

2
$\begin{array}{r} 9 \\ \times\ 3 \\ \hline \end{array}$

3
$\begin{array}{r} 0 \\ \times\ 0 \\ \hline \end{array}$

4
$\begin{array}{r} 10 \\ \times 12 \\ \hline \end{array}$

5
$\begin{array}{r} 0 \\ \times\ 1 \\ \hline \end{array}$

6
$\begin{array}{r} 4 \\ \times 11 \\ \hline \end{array}$

7
$\begin{array}{r} 2 \\ \times 11 \\ \hline \end{array}$

8
$\begin{array}{r} 9 \\ \times\ 5 \\ \hline \end{array}$

9
$\begin{array}{r} 8 \\ \times 12 \\ \hline \end{array}$

10
$\begin{array}{r} 11 \\ \times\ 4 \\ \hline \end{array}$

11
$\begin{array}{r} 12 \\ \times\ 5 \\ \hline \end{array}$

12
$\begin{array}{r} 11 \\ \times\ 3 \\ \hline \end{array}$

13
$\begin{array}{r} 7 \\ \times 12 \\ \hline \end{array}$

14
$\begin{array}{r} 9 \\ \times\ 7 \\ \hline \end{array}$

15
$\begin{array}{r} 1 \\ \times 10 \\ \hline \end{array}$

16
$\begin{array}{r} 6 \\ \times\ 8 \\ \hline \end{array}$

17
$\begin{array}{r} 10 \\ \times 11 \\ \hline \end{array}$

18
$\begin{array}{r} 7 \\ \times\ 3 \\ \hline \end{array}$

19
$\begin{array}{r} 7 \\ \times 11 \\ \hline \end{array}$

20
$\begin{array}{r} 2 \\ \times\ 8 \\ \hline \end{array}$

21
$\begin{array}{r} 3 \\ \times\ 8 \\ \hline \end{array}$

22
$\begin{array}{r} 10 \\ \times\ 7 \\ \hline \end{array}$

23
$\begin{array}{r} 8 \\ \times 10 \\ \hline \end{array}$

24
$\begin{array}{r} 5 \\ \times\ 4 \\ \hline \end{array}$

25
$\begin{array}{r} 11 \\ \times\ 1 \\ \hline \end{array}$

26
$\begin{array}{r} 5 \\ \times\ 3 \\ \hline \end{array}$

27
$\begin{array}{r} 7 \\ \times\ 6 \\ \hline \end{array}$

28
$\begin{array}{r} 8 \\ \times\ 0 \\ \hline \end{array}$

29
$\begin{array}{r} 3 \\ \times\ 5 \\ \hline \end{array}$

30
$\begin{array}{r} 9 \\ \times 10 \\ \hline \end{array}$

31
$\begin{array}{r} 3 \\ \times\ 2 \\ \hline \end{array}$

32
$\begin{array}{r} 10 \\ \times 10 \\ \hline \end{array}$

33
$\begin{array}{r} 8 \\ \times\ 2 \\ \hline \end{array}$

34
$\begin{array}{r} 11 \\ \times\ 8 \\ \hline \end{array}$

35
$\begin{array}{r} 7 \\ \times\ 8 \\ \hline \end{array}$

36
$\begin{array}{r} 11 \\ \times 10 \\ \hline \end{array}$

37
$\begin{array}{r} 11 \\ \times\ 7 \\ \hline \end{array}$

38
$\begin{array}{r} 5 \\ \times\ 9 \\ \hline \end{array}$

39
$\begin{array}{r} 10 \\ \times\ 8 \\ \hline \end{array}$

40
$\begin{array}{r} 5 \\ \times\ 5 \\ \hline \end{array}$

41
$\begin{array}{r} 10 \\ \times\ 3 \\ \hline \end{array}$

42
$\begin{array}{r} 4 \\ \times\ 8 \\ \hline \end{array}$

43
$\begin{array}{r} 12 \\ \times\ 2 \\ \hline \end{array}$

44
$\begin{array}{r} 11 \\ \times 12 \\ \hline \end{array}$

45
$\begin{array}{r} 0 \\ \times\ 2 \\ \hline \end{array}$

46
$\begin{array}{r} 6 \\ \times 10 \\ \hline \end{array}$

47
$\begin{array}{r} 3 \\ \times 10 \\ \hline \end{array}$

48
$\begin{array}{r} 8 \\ \times\ 6 \\ \hline \end{array}$

49
$\begin{array}{r} 8 \\ \times\ 7 \\ \hline \end{array}$

50
$\begin{array}{r} 6 \\ \times\ 4 \\ \hline \end{array}$

51
$\begin{array}{r} 8 \\ \times 11 \\ \hline \end{array}$

52
$\begin{array}{r} 5 \\ \times 12 \\ \hline \end{array}$

53
$\begin{array}{r} 2 \\ \times\ 2 \\ \hline \end{array}$

54
$\begin{array}{r} 3 \\ \times\ 9 \\ \hline \end{array}$

55
$\begin{array}{r} 9 \\ \times\ 2 \\ \hline \end{array}$

56
$\begin{array}{r} 8 \\ \times\ 5 \\ \hline \end{array}$

57
$\begin{array}{r} 5 \\ \times\ 2 \\ \hline \end{array}$

58
$\begin{array}{r} 5 \\ \times\ 6 \\ \hline \end{array}$

59
$\begin{array}{r} 6 \\ \times 11 \\ \hline \end{array}$

60
$\begin{array}{r} 6 \\ \times\ 6 \\ \hline \end{array}$

© 2024 Teach Elementals, LLC

MIXED REVIEW: ALL FACTS, SET P

Name: _____

Date: _____

SCORE

1	2	3	4
5 × 7	5 × 4	12 × 7	5 × 8

5	6	7	8	9	10	11	12
11 × 4	9 × 8	3 × 7	8 ×12	4 ×12	12 × 6	10 × 4	12 × 4

13	14	15	16	17	18	19	20
11 ×10	11 × 9	8 × 8	4 × 2	0 × 5	5 × 5	1 × 8	6 × 7

21	22	23	24	25	26	27	28
3 ×11	7 × 7	10 × 7	3 ×12	12 × 8	7 × 9	7 × 3	8 × 5

29	30	31	32	33	34	35	36
12 ×10	12 × 0	9 ×10	11 ×11	8 × 6	8 × 4	12 ×11	6 × 4

37	38	39	40	41	42	43	44
2 ×10	3 ×10	7 × 1	6 ×12	5 ×10	5 × 2	9 × 9	10 ×12

45	46	47	48	49	50	51	52
3 × 4	10 × 8	5 ×11	9 × 6	7 ×11	3 × 2	2 × 3	12 × 9

53	54	55	56	57	58	59	60
4 × 3	7 × 8	2 × 1	6 ×10	11 × 5	3 × 9	0 × 1	11 × 3

© 2024 Teach Elementals, LLC

Name: _____ Date: _____

SCORE

1 $\begin{array}{r} 9 \\ \times\ 8 \\ \hline \end{array}$	**2** $\begin{array}{r} 11 \\ \times\ 4 \\ \hline \end{array}$	**3** $\begin{array}{r} 3 \\ \times\ 2 \\ \hline \end{array}$	**4** $\begin{array}{r} 9 \\ \times\ 2 \\ \hline \end{array}$

5 $\begin{array}{r} 5 \\ \times\ 9 \\ \hline \end{array}$	**6** $\begin{array}{r} 12 \\ \times\ 1 \\ \hline \end{array}$	**7** $\begin{array}{r} 7 \\ \times\ 4 \\ \hline \end{array}$	**8** $\begin{array}{r} 8 \\ \times 12 \\ \hline \end{array}$	**9** $\begin{array}{r} 9 \\ \times\ 3 \\ \hline \end{array}$	**10** $\begin{array}{r} 8 \\ \times\ 0 \\ \hline \end{array}$	**11** $\begin{array}{r} 3 \\ \times 11 \\ \hline \end{array}$	**12** $\begin{array}{r} 9 \\ \times 11 \\ \hline \end{array}$
13 $\begin{array}{r} 11 \\ \times\ 2 \\ \hline \end{array}$	**14** $\begin{array}{r} 10 \\ \times\ 3 \\ \hline \end{array}$	**15** $\begin{array}{r} 6 \\ \times\ 8 \\ \hline \end{array}$	**16** $\begin{array}{r} 7 \\ \times\ 2 \\ \hline \end{array}$	**17** $\begin{array}{r} 11 \\ \times\ 8 \\ \hline \end{array}$	**18** $\begin{array}{r} 8 \\ \times\ 9 \\ \hline \end{array}$	**19** $\begin{array}{r} 6 \\ \times\ 6 \\ \hline \end{array}$	**20** $\begin{array}{r} 0 \\ \times 12 \\ \hline \end{array}$
21 $\begin{array}{r} 10 \\ \times 10 \\ \hline \end{array}$	**22** $\begin{array}{r} 12 \\ \times\ 9 \\ \hline \end{array}$	**23** $\begin{array}{r} 7 \\ \times 10 \\ \hline \end{array}$	**24** $\begin{array}{r} 4 \\ \times\ 5 \\ \hline \end{array}$	**25** $\begin{array}{r} 12 \\ \times\ 2 \\ \hline \end{array}$	**26** $\begin{array}{r} 7 \\ \times 12 \\ \hline \end{array}$	**27** $\begin{array}{r} 3 \\ \times\ 5 \\ \hline \end{array}$	**28** $\begin{array}{r} 10 \\ \times\ 6 \\ \hline \end{array}$
29 $\begin{array}{r} 8 \\ \times\ 1 \\ \hline \end{array}$	**30** $\begin{array}{r} 8 \\ \times\ 8 \\ \hline \end{array}$	**31** $\begin{array}{r} 3 \\ \times\ 3 \\ \hline \end{array}$	**32** $\begin{array}{r} 9 \\ \times\ 7 \\ \hline \end{array}$	**33** $\begin{array}{r} 12 \\ \times\ 3 \\ \hline \end{array}$	**34** $\begin{array}{r} 4 \\ \times 11 \\ \hline \end{array}$	**35** $\begin{array}{r} 8 \\ \times\ 3 \\ \hline \end{array}$	**36** $\begin{array}{r} 9 \\ \times\ 4 \\ \hline \end{array}$
37 $\begin{array}{r} 1 \\ \times\ 1 \\ \hline \end{array}$	**38** $\begin{array}{r} 5 \\ \times 11 \\ \hline \end{array}$	**39** $\begin{array}{r} 8 \\ \times 11 \\ \hline \end{array}$	**40** $\begin{array}{r} 7 \\ \times 11 \\ \hline \end{array}$	**41** $\begin{array}{r} 7 \\ \times\ 6 \\ \hline \end{array}$	**42** $\begin{array}{r} 12 \\ \times 12 \\ \hline \end{array}$	**43** $\begin{array}{r} 12 \\ \times\ 4 \\ \hline \end{array}$	**44** $\begin{array}{r} 0 \\ \times\ 6 \\ \hline \end{array}$
45 $\begin{array}{r} 10 \\ \times\ 9 \\ \hline \end{array}$	**46** $\begin{array}{r} 5 \\ \times\ 6 \\ \hline \end{array}$	**47** $\begin{array}{r} 8 \\ \times\ 5 \\ \hline \end{array}$	**48** $\begin{array}{r} 11 \\ \times\ 5 \\ \hline \end{array}$	**49** $\begin{array}{r} 10 \\ \times\ 7 \\ \hline \end{array}$	**50** $\begin{array}{r} 12 \\ \times\ 5 \\ \hline \end{array}$	**51** $\begin{array}{r} 5 \\ \times 10 \\ \hline \end{array}$	**52** $\begin{array}{r} 2 \\ \times\ 6 \\ \hline \end{array}$
53 $\begin{array}{r} 9 \\ \times\ 5 \\ \hline \end{array}$	**54** $\begin{array}{r} 8 \\ \times\ 6 \\ \hline \end{array}$	**55** $\begin{array}{r} 2 \\ \times 11 \\ \hline \end{array}$	**56** $\begin{array}{r} 3 \\ \times\ 8 \\ \hline \end{array}$	**57** $\begin{array}{r} 8 \\ \times\ 2 \\ \hline \end{array}$	**58** $\begin{array}{r} 6 \\ \times\ 5 \\ \hline \end{array}$	**59** $\begin{array}{r} 10 \\ \times\ 4 \\ \hline \end{array}$	**60** $\begin{array}{r} 4 \\ \times\ 7 \\ \hline \end{array}$

© 2024 Teach Elementals, LLC

Name: _____   Date: _____

SCORE

1 $\begin{array}{r} 9 \\ \times\ 7 \\ \hline \end{array}$	**2** $\begin{array}{r} 5 \\ \times\ 5 \\ \hline \end{array}$	**3** $\begin{array}{r} 5 \\ \times\ 2 \\ \hline \end{array}$	**4** $\begin{array}{r} 8 \\ \times\ 3 \\ \hline \end{array}$

5 $\begin{array}{r} 12 \\ \times\ 7 \\ \hline \end{array}$	**6** $\begin{array}{r} 11 \\ \times 10 \\ \hline \end{array}$	**7** $\begin{array}{r} 10 \\ \times\ 8 \\ \hline \end{array}$	**8** $\begin{array}{r} 4 \\ \times\ 4 \\ \hline \end{array}$	**9** $\begin{array}{r} 9 \\ \times\ 4 \\ \hline \end{array}$	**10** $\begin{array}{r} 11 \\ \times\ 8 \\ \hline \end{array}$	**11** $\begin{array}{r} 3 \\ \times\ 6 \\ \hline \end{array}$	**12** $\begin{array}{r} 6 \\ \times 11 \\ \hline \end{array}$
13 $\begin{array}{r} 7 \\ \times\ 7 \\ \hline \end{array}$	**14** $\begin{array}{r} 11 \\ \times\ 9 \\ \hline \end{array}$	**15** $\begin{array}{r} 3 \\ \times\ 3 \\ \hline \end{array}$	**16** $\begin{array}{r} 11 \\ \times\ 6 \\ \hline \end{array}$	**17** $\begin{array}{r} 4 \\ \times 10 \\ \hline \end{array}$	**18** $\begin{array}{r} 11 \\ \times\ 7 \\ \hline \end{array}$	**19** $\begin{array}{r} 10 \\ \times 11 \\ \hline \end{array}$	**20** $\begin{array}{r} 9 \\ \times 12 \\ \hline \end{array}$
21 $\begin{array}{r} 7 \\ \times\ 2 \\ \hline \end{array}$	**22** $\begin{array}{r} 12 \\ \times 12 \\ \hline \end{array}$	**23** $\begin{array}{r} 10 \\ \times\ 5 \\ \hline \end{array}$	**24** $\begin{array}{r} 10 \\ \times\ 9 \\ \hline \end{array}$	**25** $\begin{array}{r} 2 \\ \times\ 5 \\ \hline \end{array}$	**26** $\begin{array}{r} 8 \\ \times\ 2 \\ \hline \end{array}$	**27** $\begin{array}{r} 4 \\ \times\ 9 \\ \hline \end{array}$	**28** $\begin{array}{r} 4 \\ \times\ 2 \\ \hline \end{array}$
29 $\begin{array}{r} 5 \\ \times\ 3 \\ \hline \end{array}$	**30** $\begin{array}{r} 6 \\ \times\ 3 \\ \hline \end{array}$	**31** $\begin{array}{r} 7 \\ \times\ 3 \\ \hline \end{array}$	**32** $\begin{array}{r} 3 \\ \times\ 7 \\ \hline \end{array}$	**33** $\begin{array}{r} 1 \\ \times 10 \\ \hline \end{array}$	**34** $\begin{array}{r} 3 \\ \times\ 1 \\ \hline \end{array}$	**35** $\begin{array}{r} 7 \\ \times\ 6 \\ \hline \end{array}$	**36** $\begin{array}{r} 12 \\ \times\ 8 \\ \hline \end{array}$
37 $\begin{array}{r} 8 \\ \times\ 7 \\ \hline \end{array}$	**38** $\begin{array}{r} 4 \\ \times\ 1 \\ \hline \end{array}$	**39** $\begin{array}{r} 0 \\ \times 11 \\ \hline \end{array}$	**40** $\begin{array}{r} 6 \\ \times 10 \\ \hline \end{array}$	**41** $\begin{array}{r} 8 \\ \times\ 9 \\ \hline \end{array}$	**42** $\begin{array}{r} 5 \\ \times 12 \\ \hline \end{array}$	**43** $\begin{array}{r} 6 \\ \times\ 8 \\ \hline \end{array}$	**44** $\begin{array}{r} 3 \\ \times\ 5 \\ \hline \end{array}$
45 $\begin{array}{r} 6 \\ \times\ 9 \\ \hline \end{array}$	**46** $\begin{array}{r} 5 \\ \times\ 9 \\ \hline \end{array}$	**47** $\begin{array}{r} 12 \\ \times\ 6 \\ \hline \end{array}$	**48** $\begin{array}{r} 11 \\ \times 11 \\ \hline \end{array}$	**49** $\begin{array}{r} 3 \\ \times\ 8 \\ \hline \end{array}$	**50** $\begin{array}{r} 0 \\ \times\ 4 \\ \hline \end{array}$	**51** $\begin{array}{r} 10 \\ \times\ 2 \\ \hline \end{array}$	**52** $\begin{array}{r} 6 \\ \times\ 2 \\ \hline \end{array}$
53 $\begin{array}{r} 10 \\ \times\ 3 \\ \hline \end{array}$	**54** $\begin{array}{r} 5 \\ \times\ 7 \\ \hline \end{array}$	**55** $\begin{array}{r} 5 \\ \times\ 8 \\ \hline \end{array}$	**56** $\begin{array}{r} 7 \\ \times\ 9 \\ \hline \end{array}$	**57** $\begin{array}{r} 11 \\ \times 12 \\ \hline \end{array}$	**58** $\begin{array}{r} 1 \\ \times\ 0 \\ \hline \end{array}$	**59** $\begin{array}{r} 7 \\ \times\ 5 \\ \hline \end{array}$	**60** $\begin{array}{r} 8 \\ \times 10 \\ \hline \end{array}$

© 2024 Teach Elementals, LLC

Name: _____

Date: _____

SCORE

1 2 ×10	**2** 10 × 4	**3** 3 × 2	**4** 1 × 2
5 10 ×12	**6** 6 ×10	**7** 0 × 1	**8** 7 × 7

9 8 ×10	**10** 9 × 6	**11** 10 × 2	**12** 11 × 5
13 5 ×11	**14** 5 × 6	**15** 8 × 9	**16** 8 × 4
17 11 ×11	**18** 3 ×11	**19** 6 × 8	**20** 10 × 7
21 12 × 9	**22** 5 × 3	**23** 9 × 2	**24** 6 × 4
25 4 × 8	**26** 2 × 9	**27** 5 × 9	**28** 11 ×12
29 12 × 4	**30** 10 × 8	**31** 11 × 6	**32** 10 ×11
33 3 ×10	**34** 9 ×12	**35** 8 × 3	**36** 5 ×10
37 3 × 3	**38** 4 × 3	**39** 11 × 9	**40** 12 × 7
41 3 × 7	**42** 11 × 2	**43** 7 × 5	**44** 10 ×10
45 6 × 9	**46** 3 × 5	**47** 6 ×12	**48** 11 × 4
49 7 × 8	**50** 5 × 0	**51** 6 × 1	**52** 8 × 8
53 2 × 3	**54** 6 × 2	**55** 8 ×12	**56** 0 × 7
57 0 × 0	**58** 9 ×10	**59** 11 × 8	**60** 8 ×11

© 2024 Teach Elementals, LLC

Name: _____

Date: _____

SCORE

1 12 ×12	**2** 9 ×12	**3** 3 × 6	**4** 8 × 2

5 3 ×10	**6** 6 × 7	**7** 5 × 8	**8** 12 × 9	**9** 2 ×12	**10** 9 × 4	**11** 12 × 2	**12** 7 ×10
13 7 × 2	**14** 11 ×11	**15** 11 × 1	**16** 3 × 9	**17** 2 × 8	**18** 3 × 7	**19** 7 × 6	**20** 6 ×11
21 8 × 7	**22** 7 × 4	**23** 8 ×11	**24** 6 × 5	**25** 7 × 8	**26** 12 ×10	**27** 5 × 1	**28** 5 ×12
29 10 × 7	**30** 3 × 5	**31** 10 × 9	**32** 2 × 7	**33** 9 × 5	**34** 3 × 0	**35** 6 × 8	**36** 5 ×11
37 12 × 6	**38** 7 × 9	**39** 10 ×12	**40** 10 × 5	**41** 9 × 3	**42** 4 ×12	**43** 7 × 7	**44** 3 × 4
45 1 × 9	**46** 8 × 3	**47** 11 × 7	**48** 5 × 7	**49** 10 × 6	**50** 3 × 3	**51** 11 × 5	**52** 0 × 9
53 2 × 2	**54** 7 × 3	**55** 12 × 5	**56** 11 ×10	**57** 3 × 8	**58** 6 × 3	**59** 0 ×10	**60** 8 × 6

© 2024 Teach Elementals, LLC

Name: _____   Date: _____

SCORE

1
```
   2
×  0
```

2
```
   0
×  7
```

3
```
   4
×  6
```

4
```
   9
×  8
```

5
```
  11
×  8
```

6
```
   5
×  8
```

7
```
  10
×  3
```

8
```
  10
×  5
```

9
```
   9
×  5
```

10
```
   9
×  2
```

11
```
   0
×  6
```

12
```
   4
×  9
```

13
```
  10
× 10
```

14
```
   6
×  7
```

15
```
   9
×  9
```

16
```
   8
×  5
```

17
```
   7
×  9
```

18
```
   1
×  8
```

19
```
   9
× 11
```

20
```
   5
×  9
```

21
```
   5
×  7
```

22
```
   9
×  6
```

23
```
  12
×  6
```

24
```
  11
× 10
```

25
```
   8
×  2
```

26
```
  12
×  8
```

27
```
   6
×  1
```

28
```
   7
× 11
```

29
```
   2
×  3
```

30
```
  11
×  4
```

31
```
   3
×  4
```

32
```
  12
×  3
```

33
```
   4
× 10
```

34
```
   7
×  4
```

35
```
   5
×  5
```

36
```
  10
× 11
```

37
```
  11
×  9
```

38
```
   5
× 12
```

39
```
   7
×  6
```

40
```
   6
×  4
```

41
```
   4
×  1
```

42
```
  11
×  3
```

43
```
  12
×  7
```

44
```
   5
×  2
```

45
```
   5
×  4
```

46
```
   6
×  9
```

47
```
   6
×  6
```

48
```
   3
× 12
```

49
```
   7
× 12
```

50
```
   6
× 10
```

51
```
  12
×  5
```

52
```
   8
×  9
```

53
```
   4
×  3
```

54
```
   7
× 10
```

55
```
  10
×  9
```

56
```
  12
× 11
```

57
```
   5
×  3
```

58
```
  11
×  2
```

59
```
  11
×  6
```

60
```
   9
×  7
```

© 2024 Teach Elementals, LLC

MIXED REVIEW: ALL FACTS, SET V

Name: _____

Date: _____

SCORE

1	2	3	4
9 × 4	5 × 5	10 ×10	12 ×10

5	6	7	8	9	10	11	12
7 × 9	12 × 3	0 × 8	8 ×12	11 ×12	3 × 6	8 × 5	4 × 8

13	14	15	16	17	18	19	20
6 ×12	5 ×11	10 × 5	2 ×10	3 × 7	6 × 9	10 × 9	3 ×11

21	22	23	24	25	26	27	28
7 × 7	9 × 1	0 ×12	10 ×11	12 × 9	3 × 4	6 × 5	3 × 8

29	30	31	32	33	34	35	36
12 × 7	7 × 5	9 ×11	12 × 4	4 × 5	6 ×11	3 × 2	2 × 5

37	38	39	40	41	42	43	44
5 × 8	11 ×11	12 × 2	5 × 3	4 ×11	1 × 0	7 × 2	8 ×11

45	46	47	48	49	50	51	52
11 × 9	12 ×12	9 × 2	9 × 3	8 × 7	12 ×11	0 × 1	5 × 2

53	54	55	56	57	58	59	60
10 × 6	8 ×10	11 × 4	11 × 2	9 ×10	12 × 5	5 ×12	1 × 3

© 2024 Teach Elementals, LLC

Name: _____

Date: _____

SCORE

1
8
× 5

2
0
× 9

3
7
× 4

4
1
× 2

5
12
×12

6
10
× 7

7
8
× 8

8
6
× 9

9
7
×10

10
8
× 9

11
9
× 6

12
10
× 1

13
5
× 9

14
5
× 0

15
8
× 3

16
7
× 8

17
3
× 2

18
2
× 9

19
12
× 2

20
9
× 3

21
8
× 2

22
12
× 1

23
11
× 8

24
9
× 5

25
6
×10

26
11
× 3

27
11
× 4

28
6
× 8

29
3
×11

30
12
× 8

31
5
× 4

32
2
× 4

33
9
× 9

34
5
× 7

35
7
×11

36
8
× 4

37
2
× 6

38
10
×12

39
5
× 2

40
6
× 5

41
11
×11

42
2
× 2

43
7
× 6

44
11
× 7

45
6
× 4

46
5
× 6

47
3
×10

48
12
× 9

49
6
× 2

50
0
× 4

51
10
× 8

52
5
× 3

53
11
× 6

54
9
× 7

55
11
× 2

56
11
×10

57
9
×10

58
8
×12

59
11
× 5

60
4
×12

MIXED REVIEW: ALL FACTS, SET X

Name: _____

Date: _____

SCORE

1
10
× 3

2
3
× 5

3
7
× 6

4
3
×12

5
11
× 1

6
8
× 3

7
4
× 2

8
3
× 6

9
7
× 3

10
0
× 3

11
2
× 8

12
1
× 1

13
3
× 9

14
7
× 8

15
6
× 6

16
10
× 4

17
5
× 6

18
5
× 5

19
5
× 4

20
1
× 7

21
3
×10

22
8
× 6

23
2
×12

24
12
× 5

25
11
× 8

26
9
× 6

27
4
× 7

28
10
× 0

29
9
× 4

30
10
× 9

31
11
× 7

32
9
×12

33
5
× 9

34
5
×10

35
8
× 7

36
10
× 2

37
11
× 6

38
6
× 7

39
9
× 8

40
7
× 5

41
12
× 4

42
8
×11

43
3
× 3

44
8
× 2

45
12
×10

46
6
× 3

47
10
× 5

48
3
× 8

49
5
× 7

50
11
× 9

51
12
× 6

52
10
×11

53
0
×11

54
9
× 7

55
7
×12

56
10
× 7

57
9
× 9

58
4
× 4

59
8
×10

60
10
×10

© 2024 Teach Elementals, LLC

Name: _____   Date: _____

SCORE

1 $\begin{array}{r} 2 \\ \times\ 2 \\ \hline \end{array}$	**2** $\begin{array}{r} 0 \\ \times\ 1 \\ \hline \end{array}$	**3** $\begin{array}{r} 11 \\ \times\ 6 \\ \hline \end{array}$	**4** $\begin{array}{r} 7 \\ \times\ 3 \\ \hline \end{array}$

5 $\begin{array}{r} 2 \\ \times 10 \\ \hline \end{array}$	**6** $\begin{array}{r} 9 \\ \times 11 \\ \hline \end{array}$	**7** $\begin{array}{r} 4 \\ \times\ 1 \\ \hline \end{array}$	**8** $\begin{array}{r} 2 \\ \times 11 \\ \hline \end{array}$	**9** $\begin{array}{r} 1 \\ \times\ 5 \\ \hline \end{array}$	**10** $\begin{array}{r} 3 \\ \times\ 4 \\ \hline \end{array}$	**11** $\begin{array}{r} 6 \\ \times\ 8 \\ \hline \end{array}$	**12** $\begin{array}{r} 3 \\ \times\ 5 \\ \hline \end{array}$
13 $\begin{array}{r} 11 \\ \times 12 \\ \hline \end{array}$	**14** $\begin{array}{r} 3 \\ \times 11 \\ \hline \end{array}$	**15** $\begin{array}{r} 8 \\ \times\ 7 \\ \hline \end{array}$	**16** $\begin{array}{r} 10 \\ \times\ 8 \\ \hline \end{array}$	**17** $\begin{array}{r} 9 \\ \times\ 7 \\ \hline \end{array}$	**18** $\begin{array}{r} 2 \\ \times\ 0 \\ \hline \end{array}$	**19** $\begin{array}{r} 12 \\ \times\ 3 \\ \hline \end{array}$	**20** $\begin{array}{r} 10 \\ \times\ 6 \\ \hline \end{array}$
21 $\begin{array}{r} 0 \\ \times\ 0 \\ \hline \end{array}$	**22** $\begin{array}{r} 4 \\ \times 12 \\ \hline \end{array}$	**23** $\begin{array}{r} 5 \\ \times\ 8 \\ \hline \end{array}$	**24** $\begin{array}{r} 6 \\ \times\ 2 \\ \hline \end{array}$	**25** $\begin{array}{r} 2 \\ \times\ 5 \\ \hline \end{array}$	**26** $\begin{array}{r} 8 \\ \times\ 3 \\ \hline \end{array}$	**27** $\begin{array}{r} 2 \\ \times\ 5 \\ \hline \end{array}$	**28** $\begin{array}{r} 3 \\ \times 10 \\ \hline \end{array}$
29 $\begin{array}{r} 5 \\ \times\ 6 \\ \hline \end{array}$	**30** $\begin{array}{r} 12 \\ \times\ 2 \\ \hline \end{array}$	**31** $\begin{array}{r} 6 \\ \times\ 9 \\ \hline \end{array}$	**32** $\begin{array}{r} 9 \\ \times\ 4 \\ \hline \end{array}$	**33** $\begin{array}{r} 4 \\ \times 10 \\ \hline \end{array}$	**34** $\begin{array}{r} 10 \\ \times 11 \\ \hline \end{array}$	**35** $\begin{array}{r} 2 \\ \times\ 8 \\ \hline \end{array}$	**36** $\begin{array}{r} 9 \\ \times\ 3 \\ \hline \end{array}$
37 $\begin{array}{r} 2 \\ \times\ 6 \\ \hline \end{array}$	**38** $\begin{array}{r} 6 \\ \times 11 \\ \hline \end{array}$	**39** $\begin{array}{r} 4 \\ \times 11 \\ \hline \end{array}$	**40** $\begin{array}{r} 4 \\ \times 10 \\ \hline \end{array}$	**41** $\begin{array}{r} 10 \\ \times\ 9 \\ \hline \end{array}$	**42** $\begin{array}{r} 12 \\ \times\ 8 \\ \hline \end{array}$	**43** $\begin{array}{r} 7 \\ \times 10 \\ \hline \end{array}$	**44** $\begin{array}{r} 6 \\ \times 10 \\ \hline \end{array}$
45 $\begin{array}{r} 9 \\ \times\ 5 \\ \hline \end{array}$	**46** $\begin{array}{r} 7 \\ \times\ 1 \\ \hline \end{array}$	**47** $\begin{array}{r} 11 \\ \times\ 7 \\ \hline \end{array}$	**48** $\begin{array}{r} 7 \\ \times\ 5 \\ \hline \end{array}$	**49** $\begin{array}{r} 4 \\ \times\ 8 \\ \hline \end{array}$	**50** $\begin{array}{r} 9 \\ \times\ 9 \\ \hline \end{array}$	**51** $\begin{array}{r} 7 \\ \times 11 \\ \hline \end{array}$	**52** $\begin{array}{r} 3 \\ \times\ 9 \\ \hline \end{array}$
53 $\begin{array}{r} 10 \\ \times\ 5 \\ \hline \end{array}$	**54** $\begin{array}{r} 12 \\ \times\ 6 \\ \hline \end{array}$	**55** $\begin{array}{r} 4 \\ \times\ 6 \\ \hline \end{array}$	**56** $\begin{array}{r} 3 \\ \times\ 8 \\ \hline \end{array}$	**57** $\begin{array}{r} 5 \\ \times 11 \\ \hline \end{array}$	**58** $\begin{array}{r} 8 \\ \times 12 \\ \hline \end{array}$	**59** $\begin{array}{r} 2 \\ \times\ 6 \\ \hline \end{array}$	**60** $\begin{array}{r} 11 \\ \times\ 4 \\ \hline \end{array}$

Name: _____

Date: _____

SCORE

1 2 × 2	**2** 9 × 6	**3** 11 × 3	**4** 4 × 8
5 11 × 5	**6** 10 × 3	**7** 8 × 6	**8** 3 × 6
9 8 × 4	**10** 2 ×12	**11** 5 ×12	**12** 11 × 8
13 4 × 7	**14** 2 × 7	**15** 0 × 3	**16** 10 ×10
17 8 ×11	**18** 10 × 7	**19** 7 × 4	**20** 12 ×11
21 10 × 2	**22** 2 × 4	**23** 0 ×10	**24** 12 ×10
25 4 × 4	**26** 9 ×12	**27** 4 × 5	**28** 8 ×10
29 4 × 4	**30** 11 × 2	**31** 10 × 4	**32** 4 ×11
33 4 × 5	**34** 11 ×11	**35** 12 × 5	**36** 6 × 7
37 2 × 3	**38** 1 × 5	**39** 6 × 4	**40** 7 × 8
41 5 × 0	**42** 7 ×12	**43** 8 × 9	**44** 3 × 1
45 7 × 2	**46** 3 × 7	**47** 8 × 5	**48** 4 ×12
49 3 × 2	**50** 6 × 6	**51** 6 ×12	**52** 5 × 7
53 7 × 7	**54** 11 × 1	**55** 7 × 9	**56** 3 × 3
57 2 × 7	**58** 11 ×10	**59** 3 ×12	**60** 5 × 2

© 2024 Teach Elementals, LLC

ANSWER KEY

	Day 1	Day 2	Day 3	Day 4	Day 5	Day 6	Day 7	Day 8	Day 9	Day 10
1)	0	0	11	5	10	6	0	15	12	32
2)	0	0	6	6	20	20	24	6	33	20
3)	0	0	12	2	24	18	12	24	3	36
4)	0	0	7	9	14	16	3	21	27	28
5)	0	0	3	2	6	20	21	36	3	28
6)	0	0	9	5	24	14	6	3	15	24
7)	0	0	10	4	16	10	27	12	27	12
8)	0	0	11	7	18	24	15	27	30	16
9)	0	0	4	7	22	14	18	33	36	24
10)	0	0	5	6	8	8	18	18	36	44
11)	0	0	0	11	14	8	6	15	21	40
12)	0	0	1	0	12	12	15	6	24	0
13)	0	0	12	3	0	18	0	3	12	12
14)	0	0	10	12	18	22	33	30	18	8
15)	0	0	7	12	4	22	33	27	9	36
16)	0	0	2	11	2	4	27	21	0	48
17)	0	0	8	4	8	2	24	18	30	48
18)	0	0	9	8	0	16	30	24	6	32
19)	0	0	8	10	22	0	9	30	6	4
20)	0	0	0	0	12	2	36	0	24	8
21)	0	0	4	8	6	10	12	36	33	0
22)	0	0	2	9	20	24	3	12	18	44
23)	0	0	5	1	10	0	36	33	15	4
24)	0	0	6	10	2	6	21	0	21	40
25)	0	0	3	3	16	12	30	9	0	20
26)	0	0	12	12	20	18	9	33	3	36
27)	0	0	8	8	2	22	30	9	33	0
28)	0	0	3	3	10	0	18	0	18	40
29)	0	0	6	7	0	4	30	30	30	16
30)	0	0	5	4	0	20	15	3	0	12
31)	0	0	2	10	24	14	6	27	30	0
32)	0	0	1	7	16	24	24	15	18	48
33)	0	0	12	6	16	10	36	21	15	36
34)	0	0	2	4	10	20	27	18	27	24
35)	0	0	5	11	6	2	0	24	27	32
36)	0	0	3	6	4	14	33	24	15	12
37)	0	0	11	12	14	8	3	36	0	48
38)	0	0	10	0	12	22	3	3	3	44
39)	0	0	4	11	6	6	18	15	36	8
40)	0	0	11	8	8	6	33	27	33	32
41)	0	0	0	10	14	8	15	6	21	20
42)	0	0	0	5	8	12	12	12	36	4
43)	0	0	4	9	18	2	21	0	12	8
44)	0	0	7	2	12	16	36	12	6	40
45)	0	0	9	5	20	18	12	6	6	4
46)	0	0	9	3	18	10	21	33	21	44
47)	0	0	6	9	22	0	27	21	24	28
48)	0	0	8	2	2	16	24	18	12	24
49)	0	0	7	0	24	12	6	30	24	28
50)	0	0	10	1	22	24	0	36	9	20
51)	0	0	0	10	20	24	27	3	3	40
52)	0	0	1	4	12	12	21	15	3	12
53)	0	0	3	5	0	8	15	24	9	44
54)	0	0	4	10	12	2	21	21	36	32
55)	0	0	6	2	18	2	0	33	0	0
56)	0	0	9	11	20	4	15	6	30	32
57)	0	0	11	0	4	0	33	27	12	28
58)	0	0	9	4	6	22	12	36	27	24
59)	0	0	7	11	18	14	12	15	21	36
60)	0	0	6	0	14	0	9	18	15	8

ANSWER KEY

KEY

	Day 11	Day 12	Day 13	Day 14	Day 15	Day 16	Day 17	Day 18	Day 19	Day 20
1)	28	32	0	10	60	24	66	54	24	56
2)	44	40	45	20	40	60	72	48	72	77
3)	48	44	50	60	30	12	54	60	60	77
4)	16	4	30	55	10	48	12	0	72	14
5)	40	12	15	60	0	54	6	36	54	0
6)	12	4	45	55	55	48	54	18	42	0
7)	8	0	10	50	35	18	60	18	54	63
8)	0	20	40	30	15	66	18	42	0	21
9)	8	36	40	40	5	18	48	6	12	84
10)	28	24	10	40	40	30	12	30	12	56
11)	48	48	5	35	25	30	48	0	24	21
12)	24	28	35	10	35	6	42	66	18	35
13)	0	28	60	20	55	12	24	42	66	7
14)	40	20	55	0	20	24	72	24	30	14
15)	4	44	0	15	15	72	36	66	0	84
16)	24	12	15	50	30	60	18	12	48	42
17)	32	8	20	30	20	0	66	12	66	35
18)	20	8	60	25	10	66	60	24	48	63
19)	44	0	55	45	60	42	24	54	6	42
20)	4	40	30	15	45	6	30	72	36	70
21)	12	36	20	45	0	54	42	6	30	28
22)	36	32	5	40	50	42	0	60	18	28
23)	20	24	35	5	50	36	30	30	6	70
24)	36	48	50	0	45	0	6	48	42	7
25)	32	16	25	35	5	72	0	72	60	49
26)	36	36	45	10	5	0	12	0	30	7
27)	36	24	60	15	60	42	54	18	60	42
28)	32	44	15	50	50	12	6	42	36	84
29)	32	12	35	0	45	36	36	0	42	28
30)	24	48	55	50	55	60	24	24	66	21
31)	24	12	45	35	30	54	60	60	24	84
32)	28	4	40	5	40	48	48	24	54	70
33)	40	40	15	0	10	48	42	36	24	35
34)	12	36	50	10	30	60	60	6	54	63
35)	16	24	0	55	55	54	66	6	18	70
36)	8	28	20	45	15	24	72	42	60	56
37)	48	0	0	45	45	18	42	60	30	49
38)	44	4	30	15	0	66	72	12	48	35
39)	12	44	5	55	10	0	12	54	72	56
40)	20	8	10	30	35	6	0	72	6	14
41)	0	28	20	5	40	18	18	54	48	42
42)	0	16	25	35	0	66	6	66	0	77
43)	4	32	50	60	50	72	24	72	42	21
44)	20	20	5	20	15	12	30	18	72	0
45)	8	48	40	25	60	6	66	48	0	63
46)	40	20	35	40	35	30	30	66	66	14
47)	48	0	10	40	20	42	48	30	12	7
48)	44	32	30	20	20	24	54	48	12	28
49)	28	40	60	30	25	30	18	12	6	77
50)	4	8	55	60	5	72	12	30	18	0
51)	16	4	20	35	40	60	72	66	0	56
52)	28	4	60	10	40	54	54	48	12	35
53)	20	40	35	0	15	24	72	36	60	63
54)	8	8	50	40	30	42	6	66	54	14
55)	40	20	5	45	30	6	60	6	24	42
56)	8	32	60	5	15	60	36	6	24	35
57)	4	0	55	15	45	0	60	30	36	63
58)	44	12	40	60	45	12	48	48	42	77
59)	4	36	15	10	55	66	24	24	30	28
60)	32	44	10	5	0	48	66	60	54	70

© 2024 Teach Elementals, LLC

ANSWER KEY

#	Day 21	Day 22	Day 23	Day 24	Day 25	Day 26	Day 27	Day 28	Day 29	Day 30
1)	77	63	35	0	48	56	0	90	0	72
2)	0	84	14	40	88	24	8	72	54	0
3)	77	56	63	16	8	88	40	9	0	45
4)	0	77	70	96	40	72	48	45	108	18
5)	42	56	7	24	16	32	96	108	81	27
6)	7	77	0	48	72	96	32	90	45	90
7)	21	14	70	72	72	40	80	63	90	9
8)	35	14	77	0	8	96	88	18	63	72
9)	21	28	14	40	16	80	72	18	108	27
10)	63	42	28	56	56	40	16	108	45	99
11)	14	7	84	16	0	48	24	99	36	108
12)	70	0	42	24	96	8	56	27	9	36
13)	84	63	21	8	40	16	8	54	36	90
14)	84	0	21	48	32	72	64	0	99	54
15)	14	21	56	88	48	32	72	36	27	54
16)	7	70	0	8	80	0	16	27	9	81
17)	28	28	49	56	0	56	0	45	99	0
18)	49	49	35	80	24	64	56	99	72	45
19)	63	7	77	96	64	48	48	63	18	99
20)	56	21	7	64	80	0	88	0	27	63
21)	35	35	56	88	32	8	24	54	90	108
22)	70	35	63	72	56	80	40	9	63	63
23)	56	70	84	32	88	88	80	36	54	18
24)	28	42	42	80	96	24	96	81	72	9
25)	42	84	28	32	24	16	32	72	18	36
26)	35	70	21	16	0	32	96	45	54	63
27)	70	84	7	80	32	48	40	99	27	54
28)	35	28	28	40	72	48	80	36	81	90
29)	70	35	0	56	48	16	32	18	90	9
30)	21	7	70	80	24	80	56	36	36	36
31)	28	21	56	16	64	0	32	9	0	99
32)	84	0	0	32	32	56	8	81	18	0
33)	14	49	70	48	16	40	88	27	27	18
34)	0	42	49	64	96	24	0	54	99	108
35)	49	14	77	88	80	96	16	72	99	36
36)	77	63	63	24	16	88	48	90	45	45
37)	7	28	63	72	8	72	72	90	72	0
38)	21	70	84	56	96	0	0	63	72	108
39)	84	77	56	72	88	8	16	99	90	18
40)	42	21	14	40	80	8	80	63	108	45
41)	56	56	28	48	40	72	48	0	9	27
42)	63	84	14	24	24	64	72	9	63	72
43)	7	56	21	0	8	96	40	0	63	27
44)	14	42	84	96	40	80	96	27	54	90
45)	42	35	42	8	0	40	24	18	0	81
46)	77	0	35	8	72	32	64	72	18	54
47)	56	7	77	88	56	56	8	108	9	63
48)	63	14	35	32	88	24	88	45	108	9
49)	0	77	7	0	48	88	56	54	36	99
50)	28	63	42	96	56	16	24	108	45	72
51)	56	56	35	72	48	88	64	18	63	108
52)	21	0	0	56	0	56	8	54	72	9
53)	77	49	7	0	8	48	88	0	45	27
54)	7	77	84	80	88	16	32	36	9	90
55)	7	35	84	96	16	0	56	63	18	99
56)	42	7	14	64	80	32	48	54	9	45
57)	63	28	0	16	72	0	40	63	27	18
58)	0	42	77	24	48	96	96	45	45	36
59)	63	70	70	40	0	72	0	27	90	108
60)	35	84	21	40	64	80	16	108	0	72

© 2024 Teach Elementals, LLC

ANSWER KEY

KEY

	Day 31	Day 32	Day 33	Day 34	Day 35	Day 36	Day 37	Day 38	Day 39	Day 40
1)	36	50	70	110	110	55	132	88	11	132
2)	99	10	20	10	90	110	88	77	88	108
3)	27	10	50	50	20	55	121	44	99	48
4)	72	20	60	110	10	11	44	33	110	108
5)	72	80	70	30	20	121	33	132	0	96
6)	18	90	0	40	70	0	88	0	77	96
7)	63	60	20	100	80	44	99	66	11	24
8)	27	110	30	60	40	77	99	77	55	12
9)	0	0	110	110	10	44	33	132	22	120
10)	54	80	80	20	120	11	110	55	110	60
11)	81	50	90	70	30	77	110	0	88	36
12)	63	110	10	80	100	99	132	11	55	72
13)	0	70	10	90	0	132	0	44	132	48
14)	45	30	110	20	0	99	22	110	33	144
15)	36	60	0	60	30	22	22	99	66	0
16)	45	120	80	10	40	22	44	110	33	36
17)	9	40	100	120	60	33	66	66	77	120
18)	90	120	40	30	50	66	77	33	0	12
19)	54	70	120	80	50	88	66	121	44	132
20)	99	40	40	0	110	0	0	55	66	84
21)	90	30	30	120	70	66	55	88	99	60
22)	108	90	90	90	120	88	11	99	121	0
23)	9	20	120	40	90	132	77	11	22	84
24)	108	100	60	70	60	110	11	22	132	24
25)	18	0	50	50	80	33	55	22	44	72
26)	63	20	90	110	90	55	55	77	33	0
27)	99	0	50	90	30	33	11	22	77	84
28)	63	30	60	90	20	22	77	132	55	48
29)	54	30	80	80	80	132	22	110	11	96
30)	72	120	110	80	20	0	110	55	33	24
31)	45	50	60	120	50	22	121	22	44	24
32)	0	40	50	20	110	110	132	33	22	84
33)	9	100	40	120	60	77	99	44	132	12
34)	45	40	10	0	50	44	77	33	110	132
35)	18	0	40	70	100	88	33	55	22	144
36)	54	90	20	0	10	88	55	77	0	12
37)	36	60	70	50	120	0	66	0	88	132
38)	36	20	0	100	30	33	0	110	11	36
39)	90	50	10	60	110	44	11	132	99	120
40)	18	10	120	110	10	55	44	11	66	36
41)	27	90	30	60	70	11	110	0	110	72
42)	0	10	100	40	0	110	33	66	99	96
43)	99	120	20	40	120	99	22	44	44	120
44)	90	80	110	30	0	99	132	88	77	0
45)	27	80	0	20	60	66	44	99	55	60
46)	108	70	70	70	40	11	88	11	88	108
47)	72	110	90	10	80	121	88	88	0	108
48)	81	70	120	30	40	77	0	121	66	60
49)	9	110	30	50	90	132	66	99	132	72
50)	108	60	80	10	70	66	99	66	121	48
51)	72	120	10	40	20	33	0	33	11	120
52)	63	40	80	50	10	22	77	77	99	108
53)	36	0	0	10	40	132	88	99	66	12
54)	0	100	10	0	90	44	110	44	110	36
55)	54	110	0	90	40	132	55	121	0	36
56)	81	10	70	70	50	0	11	66	22	0
57)	18	90	40	10	20	99	44	132	55	132
58)	90	50	30	50	100	77	11	33	22	24
59)	45	10	60	40	110	44	132	88	110	120
60)	27	80	20	100	50	66	88	99	55	84

© 2024 Teach Elementals, LLC

ANSWER KEY

	Day 41	Day 42	Day 43	Day 44	Day 45	Day 46	Day 47	Day 48	Day 49	Day 50
1)	12	24	84	11	6	50	80	55	77	55
2)	72	84	108	12	18	4	108	36	28	72
3)	132	120	96	10	0	12	6	24	77	16
4)	84	108	132	24	0	10	0	0	72	24
5)	108	24	72	5	6	0	22	30	63	110
6)	132	12	24	27	18	2	9	90	120	54
7)	96	0	108	22	27	45	54	7	49	16
8)	24	72	60	21	4	28	99	120	121	144
9)	36	108	24	12	0	14	84	132	60	24
10)	48	48	48	3	44	2	36	24	16	132
11)	0	96	12	15	30	0	18	66	42	0
12)	24	12	48	2	8	0	0	72	96	7
13)	36	36	0	24	9	8	48	99	8	36
14)	60	60	72	50	16	7	30	50	21	0
15)	120	60	120	8	24	60	110	60	0	66
16)	72	72	12	12	4	14	21	24	40	96
17)	96	36	60	0	2	33	77	45	84	66
18)	0	84	144	20	28	30	0	110	120	10
19)	48	120	96	55	40	0	132	7	108	99
20)	60	0	120	21	0	45	63	32	66	88
21)	144	144	84	0	9	18	96	9	63	63
22)	12	48	36	20	10	8	16	72	33	90
23)	84	132	36	12	5	0	66	12	80	35
24)	120	96	132	35	0	0	70	54	60	108
25)	108	132	0	20	40	0	96	55	70	0
26)	120	48	48	30	0	6	28	108	40	42
27)	12	60	0	32	6	6	45	72	30	18
28)	12	120	108	36	10	0	90	84	56	88
29)	132	60	84	8	7	24	0	40	30	56
30)	108	72	144	10	0	0	32	77	60	70
31)	60	36	96	0	25	22	88	6	42	56
32)	132	132	36	0	44	18	44	144	54	108
33)	84	12	132	30	16	15	48	12	63	77
34)	24	72	48	0	4	0	48	108	0	0
35)	36	96	120	0	35	0	81	48	90	72
36)	108	108	60	1	60	0	30	88	10	0
37)	48	84	132	33	32	0	54	20	56	40
38)	36	12	108	0	15	9	72	99	36	63
39)	144	120	12	0	15	0	72	35	14	12
40)	24	84	84	15	20	5	0	88	12	21
41)	96	144	12	14	24	20	12	40	120	10
42)	72	36	60	0	36	48	110	80	54	36
43)	120	48	120	48	16	36	56	0	99	120
44)	60	0	24	10	20	6	66	10	11	11
45)	96	24	0	4	30	40	24	72	70	12
46)	0	0	96	11	6	10	60	0	14	49
47)	84	108	24	0	0	1	88	35	56	72
48)	48	96	72	12	55	12	36	90	132	30
49)	0	132	72	0	0	33	50	33	48	0
50)	72	24	36	0	45	0	60	70	42	60
51)	48	12	60	36	12	4	80	96	72	72
52)	36	24	132	48	8	3	20	11	0	40
53)	72	12	36	3	14	3	24	0	18	90
54)	84	24	12	12	2	0	24	44	8	48
55)	84	132	132	5	24	0	63	48	18	36
56)	36	48	24	0	20	0	0	0	0	28
57)	24	72	120	40	4	55	18	110	100	48
58)	120	108	84	0	3	0	27	22	132	60
59)	96	96	12	0	5	8	120	27	36	42
60)	144	132	96	0	60	24	64	42	84	90

© 2024 Teach Elementals, LLC

ANSWER KEY

KEY

	Day 51	Day 52	Day 53	Day 54	Day 55	Day 56	Day 57	Day 58	Day 59	Day 60
1)	100	11	4	36	100	20	77	33	5	35
2)	120	60	0	24	0	20	1	9	21	77
3)	88	32	20	48	20	48	11	15	3	3
4)	24	20	16	0	4	0	11	45	33	27
5)	48	84	24	100	48	48	5	3	45	9
6)	72	42	72	96	36	64	21	35	35	81
7)	6	0	80	60	60	32	9	99	9	121
8)	77	48	24	60	96	60	45	63	99	7
9)	36	64	32	12	80	24	55	77	1	99
10)	18	21	0	12	0	8	21	45	55	9
11)	70	0	60	72	32	24	45	15	25	33
12)	132	36	64	0	16	0	9	33	9	15
13)	6	33	12	144	24	4	7	11	15	99
14)	60	48	48	8	48	60	35	25	55	33
15)	60	24	80	20	12	24	121	7	81	55
16)	60	0	60	40	16	48	27	3	121	63
17)	12	20	8	32	80	16	25	55	7	1
18)	7	8	48	120	60	72	77	55	63	77
19)	30	72	16	48	40	120	33	27	3	9
20)	84	22	40	64	32	8	33	21	9	5
21)	9	120	32	16	8	80	99	121	21	3
22)	0	44	120	80	120	16	81	5	77	15
23)	96	45	120	48	144	48	63	63	77	11
24)	8	132	24	20	40	72	7	27	27	21
25)	9	72	96	0	120	0	35	81	5	45
26)	66	40	12	72	96	12	99	9	35	55
27)	90	66	96	16	12	80	49	35	7	63
28)	22	70	100	40	24	40	55	7	45	21
29)	108	70	40	48	72	24	5	11	15	25
30)	80	30	72	24	48	32	15	77	27	7
31)	84	30	24	24	48	12	63	5	63	45
32)	110	110	36	4	72	120	3	21	99	49
33)	63	108	48	80	8	96	15	49	33	35
34)	0	96	48	8	16	100	9	9	11	27
35)	40	32	8	96	20	40	3	99	11	5
36)	80	35	144	24	64	16	27	1	49	11
37)	18	55	0	96	0	100	35	21	5	15
38)	44	42	20	40	96	20	77	49	9	35
39)	50	28	48	24	20	36	55	3	33	3
40)	45	24	48	8	24	32	3	9	25	7
41)	33	77	32	32	48	8	63	45	63	9
42)	99	80	16	24	24	48	55	15	99	55
43)	27	56	80	12	96	16	15	77	81	63
44)	18	63	36	96	60	24	5	77	15	15
45)	84	14	24	60	80	40	11	99	33	63
46)	48	12	100	20	32	24	99	7	9	21
47)	80	24	40	4	72	144	11	55	63	77
48)	54	27	24	48	100	48	77	55	35	45
49)	50	54	8	12	80	8	7	9	55	27
50)	72	99	60	48	4	48	21	27	5	99
51)	96	120	12	64	16	60	45	63	3	27
52)	88	132	4	36	12	60	9	45	55	45
53)	48	99	48	16	8	16	99	11	45	49
54)	77	0	0	0	0	0	7	3	27	5
55)	56	28	24	120	64	16	27	99	45	121
56)	0	14	120	48	24	96	15	27	77	1
57)	81	54	120	40	16	24	9	15	3	11
58)	35	0	32	8	32	64	1	25	77	9
59)	32	121	64	60	40	48	9	121	11	3
60)	0	110	24	16	48	32	3	5	1	33

© 2024 Teach Elementals, LLC

ANSWER KEY

	Day 61	Day 62	Day 63	Day 64	Day 65	Day 66	Day 67	Day 68	Day 69	Day 70
1)	4	4	0	72	9	9	20	5	30	48
2)	8	20	16	24	27	45	35	90	40	42
3)	24	56	96	30	36	99	60	55	50	0
4)	16	16	32	60	24	9	110	20	0	66
5)	18	20	6	36	18	36	30	90	40	72
6)	32	0	28	12	42	30	10	40	45	45
7)	64	18	56	42	3	108	15	20	35	36
8)	12	14	48	54	72	0	25	50	35	0
9)	4	44	88	12	108	15	80	50	0	144
10)	16	16	8	66	45	21	40	45	15	63
11)	24	8	16	81	0	72	50	10	55	9
12)	20	24	4	63	33	81	40	10	55	90
13)	24	24	24	33	99	18	120	15	70	30
14)	16	22	8	18	15	24	20	15	20	108
15)	8	12	40	0	60	24	90	100	20	99
16)	40	8	44	0	18	18	70	50	45	27
17)	8	6	20	54	63	12	45	40	110	60
18)	0	64	2	6	81	0	120	35	40	36
19)	96	88	72	9	12	63	60	80	5	96
20)	0	80	40	30	27	36	80	70	50	108
21)	36	28	12	36	54	6	5	25	40	60
22)	6	32	80	18	54	90	50	110	10	36
23)	10	72	32	27	30	27	0	80	30	84
24)	22	0	20	0	72	27	10	120	90	12
25)	20	2	0	6	0	36	40	60	80	6
26)	0	12	14	48	6	42	50	60	100	48
27)	72	24	8	90	90	18	70	0	50	0
28)	80	48	24	9	12	60	55	55	10	12
29)	44	16	4	108	24	54	0	35	10	54
30)	12	36	64	72	18	72	90	60	15	120
31)	28	48	0	99	30	48	55	5	50	54
32)	88	0	36	24	6	12	20	0	60	18
33)	56	40	10	45	66	54	60	50	30	24
34)	32	10	18	18	36	6	45	60	80	72
35)	40	32	12	15	36	33	15	0	60	72
36)	2	40	24	27	21	0	100	110	20	81
37)	48	8	22	21	21	30	40	120	0	18
38)	14	4	16	3	0	3	10	45	110	24
39)	48	96	48	36	9	66	0	40	60	132
40)	72	0	4	9	54	24	50	30	10	84
41)	0	88	0	18	0	9	60	30	0	72
42)	22	0	72	27	30	54	110	0	120	54
43)	24	8	10	54	60	0	30	70	25	60
44)	0	96	16	66	12	27	5	20	90	108
45)	40	8	12	6	6	3	20	30	5	0
46)	18	24	22	48	18	108	30	10	70	18
47)	36	2	0	54	12	72	30	20	120	18
48)	16	12	2	99	27	21	35	10	60	30
49)	12	8	48	90	54	18	0	30	20	36
50)	8	20	48	63	3	60	10	40	30	54
51)	10	16	8	60	36	33	40	120	120	90
52)	4	6	56	24	18	6	110	45	25	72
53)	40	12	28	45	36	6	25	20	90	99
54)	8	24	24	6	9	30	110	15	30	12
55)	88	28	96	30	99	15	90	25	10	45
56)	44	56	40	0	48	48	0	55	90	0
57)	96	32	88	12	33	54	120	60	45	63
58)	0	48	80	30	27	18	50	10	30	108
59)	24	4	44	72	42	12	60	70	0	48
60)	32	40	32	21	6	45	0	5	20	48

© 2024 Teach Elementals, LLC

ANSWER KEY

KEY

Day 71	Day 72	Day 73	Day 74	Day 75	Day 76	Day 77	Day 78	Day 79	Day 80
1) 27	1) 18	1) 14	1) 28	1) 15	1) 40	1) 66	1) 9	1) 60	1) 45
2) 72	2) 6	2) 42	2) 0	2) 70	2) 4	2) 120	2) 0	2) 12	2) 9
3) 0	3) 90	3) 15	3) 77	3) 0	3) 28	3) 55	3) 20	3) 18	3) 50
4) 120	4) 108	4) 21	4) 18	4) 8	4) 32	4) 63	4) 12	4) 24	4) 121
5) 72	5) 63	5) 14	5) 6	5) 1	5) 18	5) 36	5) 15	5) 15	5) 90
6) 48	6) 72	6) 18	6) 56	6) 96	6) 70	6) 90	6) 40	6) 8	6) 77
7) 54	7) 48	7) 0	7) 27	7) 21	7) 2	7) 144	7) 14	7) 0	7) 22
8) 36	8) 60	8) 24	8) 14	8) 77	8) 18	8) 120	8) 96	8) 110	8) 48
9) 12	9) 60	9) 12	9) 33	9) 54	9) 54	9) 80	9) 16	9) 108	9) 56
10) 24	10) 144	10) 3	10) 21	10) 90	10) 30	10) 20	10) 10	10) 100	10) 4
11) 36	11) 36	11) 35	11) 15	11) 121	11) 27	11) 9	11) 28	11) 0	11) 0
12) 45	12) 81	12) 27	12) 28	12) 55	12) 36	12) 0	12) 72	12) 32	12) 35
13) 60	13) 96	13) 84	13) 84	13) 20	13) 12	13) 6	13) 56	13) 81	13) 72
14) 72	14) 0	14) 35	14) 15	14) 6	14) 14	14) 35	14) 72	14) 9	14) 60
15) 18	15) 72	15) 6	15) 63	15) 18	15) 66	15) 50	15) 70	15) 36	15) 66
16) 144	16) 132	16) 3	16) 30	16) 50	16) 108	16) 56	16) 36	16) 50	16) 33
17) 63	17) 108	17) 0	17) 33	17) 36	17) 44	17) 96	17) 22	17) 42	17) 54
18) 42	18) 18	18) 70	18) 36	18) 42	18) 0	18) 90	18) 30	18) 108	18) 0
19) 99	19) 12	19) 21	19) 42	19) 6	19) 9	19) 20	19) 32	19) 10	19) 27
20) 12	20) 84	20) 15	20) 7	20) 96	20) 30	20) 22	20) 55	20) 132	20) 36
21) 54	21) 72	21) 12	21) 7	21) 15	21) 56	21) 36	21) 8	21) 21	21) 30
22) 108	22) 9	22) 33	22) 14	22) 60	22) 12	22) 72	22) 55	22) 5	22) 18
23) 30	23) 120	23) 27	23) 63	23) 63	23) 63	23) 0	23) 12	23) 20	23) 22
24) 90	24) 66	24) 42	24) 12	24) 28	24) 80	24) 40	24) 0	24) 15	24) 60
25) 0	25) 36	25) 63	25) 30	25) 0	25) 99	25) 4	25) 30	25) 18	25) 60
26) 6	26) 42	26) 36	26) 9	26) 14	26) 20	26) 7	26) 60	26) 30	26) 40
27) 48	27) 24	27) 70	27) 12	27) 50	27) 60	27) 40	27) 54	27) 88	27) 66
28) 0	28) 30	28) 9	28) 0	28) 81	28) 88	28) 60	28) 35	28) 50	28) 144
29) 66	29) 0	29) 77	29) 24	29) 18	29) 30	29) 45	29) 0	29) 35	29) 42
30) 36	30) 0	30) 56	30) 6	30) 72	30) 80	30) 24	30) 99	30) 40	30) 1
31) 96	31) 54	31) 63	31) 70	31) 10	31) 8	31) 96	31) 56	31) 12	31) 88
32) 84	32) 12	32) 21	32) 21	32) 88	32) 72	32) 11	32) 36	32) 80	32) 70
33) 81	33) 99	33) 18	33) 35	33) 84	33) 110	33) 21	33) 6	33) 99	33) 80
34) 18	34) 36	34) 56	34) 3	34) 36	34) 84	34) 66	34) 9	34) 35	34) 44
35) 132	35) 45	35) 6	35) 49	35) 99	35) 14	35) 110	35) 24	35) 88	35) 20
36) 24	36) 24	36) 36	36) 70	36) 72	36) 0	36) 30	36) 49	36) 72	36) 18
37) 60	37) 48	37) 21	37) 36	37) 120	37) 18	37) 36	37) 33	37) 120	37) 54
38) 9	38) 54	38) 30	38) 21	38) 27	38) 100	38) 77	38) 40	38) 25	38) 77
39) 108	39) 27	39) 49	39) 42	39) 132	39) 10	39) 33	39) 81	39) 48	39) 56
40) 0	40) 30	40) 0	40) 18	40) 18	40) 48	40) 72	40) 10	40) 27	40) 110
41) 48	41) 132	41) 24	41) 77	41) 108	41) 24	41) 32	41) 120	41) 8	41) 40
42) 0	42) 99	42) 7	42) 0	42) 24	42) 110	42) 12	42) 60	42) 21	42) 18
43) 12	43) 54	43) 28	43) 0	43) 42	43) 44	43) 45	43) 84	43) 84	43) 12
44) 18	44) 36	44) 7	44) 21	44) 132	44) 24	44) 15	44) 25	44) 55	44) 20
45) 63	45) 120	45) 33	45) 35	45) 24	45) 48	45) 54	45) 36	45) 121	45) 40
46) 72	46) 24	46) 0	46) 56	46) 32	46) 60	46) 56	46) 100	46) 48	46) 84
47) 12	47) 84	47) 30	47) 24	47) 35	47) 0	47) 88	47) 120	47) 30	47) 40
48) 66	48) 6	48) 77	48) 27	48) 0	48) 144	48) 36	48) 144	48) 18	48) 90
49) 30	49) 18	49) 28	49) 84	49) 63	49) 42	49) 24	49) 44	49) 110	49) 27
50) 9	50) 9	50) 84	50) 3	50) 20	50) 120	50) 0	50) 12	50) 96	50) 33
51) 60	51) 36	51) 30	51) 42	51) 44	51) 16	51) 64	51) 108	51) 63	51) 20
52) 6	52) 72	52) 56	52) 0	52) 18	52) 0	52) 132	52) 60	52) 132	52) 28
53) 108	53) 12	53) 35	53) 63	53) 16	53) 55	53) 132	53) 16	53) 49	53) 72
54) 24	54) 72	54) 21	54) 36	54) 22	54) 33	54) 25	54) 77	54) 80	54) 0
55) 99	55) 90	55) 56	55) 9	55) 45	55) 30	55) 88	55) 24	55) 30	55) 63
56) 132	56) 48	56) 21	56) 14	56) 12	56) 99	56) 10	56) 48	56) 90	56) 0
57) 108	57) 66	57) 6	57) 12	57) 99	57) 3	57) 36	57) 24	57) 24	57) 64
58) 72	58) 0	58) 12	58) 35	58) 24	58) 121	58) 60	58) 63	58) 90	58) 88
59) 120	59) 0	59) 14	59) 70	59) 49	59) 36	59) 15	59) 96	59) 0	59) 44
60) 60	60) 63	60) 24	60) 6	60) 60	60) 48	60) 40	60) 45	60) 42	60) 32

© 2024 Teach Elementals, LLC

ANSWER KEY

	Day 81	Day 82	Day 83	Day 84	Day 85	Day 86	Day 87	Day 88	Day 89	Day 90
1)	60	90	0	4	12	6	48	99	99	35
2)	6	0	81	70	72	27	24	14	27	20
3)	25	55	49	25	40	80	40	63	0	84
4)	110	44	80	24	40	70	132	120	120	40
5)	6	21	100	21	110	72	55	24	0	44
6)	48	0	120	36	20	16	0	21	44	72
7)	0	42	22	4	14	30	66	72	22	21
8)	36	144	110	48	30	56	24	100	45	96
9)	55	36	16	60	12	22	54	60	96	48
10)	24	80	99	132	80	27	72	18	44	72
11)	120	24	88	24	10	45	45	32	60	40
12)	56	6	24	144	9	48	36	12	33	48
13)	80	14	50	32	32	66	72	25	84	110
14)	96	30	35	0	18	42	60	20	63	99
15)	20	77	110	50	96	50	5	0	10	64
16)	90	40	30	12	0	1	56	36	48	8
17)	18	45	36	9	36	81	21	42	110	0
18)	30	63	72	60	84	72	14	21	21	25
19)	8	30	15	36	100	22	20	132	77	8
20)	81	88	72	56	15	28	36	80	16	42
21)	35	99	132	48	16	24	70	35	24	33
22)	54	9	8	18	48	14	108	84	70	49
23)	70	27	96	54	132	42	3	33	80	70
24)	4	55	70	40	0	0	40	72	20	36
25)	24	120	72	40	5	60	0	55	11	96
26)	35	6	28	55	30	21	120	36	15	63
27)	121	60	20	45	50	66	36	27	42	21
28)	21	22	108	55	24	0	8	27	0	40
29)	27	2	108	77	32	35	132	121	15	120
30)	54	21	45	8	60	36	50	6	90	0
31)	77	108	121	90	110	15	72	81	6	90
32)	12	66	20	88	60	24	12	90	100	121
33)	63	56	99	120	0	108	33	0	16	48
34)	10	60	96	0	84	99	96	12	88	32
35)	0	90	30	88	35	121	20	18	56	132
36)	66	27	24	10	48	9	55	30	110	24
37)	42	60	44	12	60	120	42	66	77	20
38)	15	12	63	99	144	16	18	16	45	30
39)	33	20	24	64	6	72	24	66	80	7
40)	30	64	8	12	0	48	30	40	25	72
41)	55	16	2	11	99	99	28	96	30	50
42)	48	70	33	63	96	48	99	36	32	10
43)	16	33	50	44	132	35	18	54	24	81
44)	18	28	14	60	72	6	30	24	132	120
45)	0	0	56	15	42	110	63	49	0	12
46)	36	18	63	70	36	20	64	0	60	80
47)	7	14	24	0	80	33	12	77	30	55
48)	48	40	60	54	42	64	144	72	48	54
49)	32	3	18	108	30	96	32	18	56	77
50)	84	60	30	63	77	54	22	40	24	6
51)	132	12	0	14	24	90	84	9	88	6
52)	56	36	24	77	70	40	14	50	60	108
53)	12	24	36	30	63	88	12	12	4	12
54)	90	132	72	22	120	30	16	1	27	56
55)	10	45	48	36	20	49	88	88	18	2
56)	36	18	18	84	77	15	0	44	40	60
57)	72	110	0	30	36	18	60	9	10	55
58)	66	15	77	50	81	6	72	64	30	27
59)	84	96	9	28	27	56	48	35	66	0
60)	24	40	7	33	24	0	54	108	36	33

© 2024 Teach Elementals, LLC

	Day 91	Day 92	Day 93	Day 94	Day 95	Day 96	Day 97	Day 98	Day 99	Day 100
1)	72	63	20	144	0	36	40	30	4	4
2)	44	25	40	108	0	25	0	15	0	54
3)	6	10	6	18	24	100	28	42	66	33
4)	18	24	2	16	72	120	2	36	21	32
5)	45	84	120	30	88	63	144	11	20	55
6)	12	110	60	42	40	36	70	24	99	30
7)	28	80	0	40	30	0	64	8	4	48
8)	96	16	49	108	50	96	54	18	22	18
9)	27	36	80	24	45	132	70	21	5	32
10)	0	88	54	36	18	18	72	0	12	24
11)	33	18	20	24	0	40	54	16	48	60
12)	99	66	55	70	36	32	10	1	15	88
13)	22	49	55	14	100	72	45	27	132	28
14)	30	99	30	121	42	55	0	56	33	14
15)	48	9	72	11	81	50	24	36	56	0
16)	14	66	32	27	40	20	56	40	80	100
17)	88	40	121	16	63	21	6	30	63	88
18)	72	77	33	21	8	54	18	25	0	70
19)	36	110	48	42	99	90	24	20	36	28
20)	0	108	70	66	45	33	27	7	60	132
21)	100	14	108	56	35	49	16	30	0	20
22)	108	144	15	28	54	9	12	48	48	8
23)	70	50	18	88	72	0	88	24	40	0
24)	20	90	24	30	110	110	45	60	12	120
25)	24	10	32	56	16	108	60	88	10	16
26)	84	16	18	120	96	12	33	54	24	108
27)	15	36	45	5	6	30	44	28	10	20
28)	60	8	132	60	77	24	48	0	30	80
29)	8	15	48	70	6	84	33	36	30	16
30)	64	18	80	15	44	35	96	90	24	22
31)	9	21	66	90	12	99	20	77	54	40
32)	63	21	110	14	36	48	8	108	36	44
33)	36	10	30	45	40	20	81	45	40	20
34)	44	3	108	0	28	66	35	50	110	121
35)	24	42	24	48	25	6	77	56	16	60
36)	36	96	50	55	110	10	32	20	27	42
37)	1	56	9	72	99	40	12	66	12	6
38)	55	4	12	63	60	121	120	42	66	5
39)	88	0	99	120	42	24	10	72	44	24
40)	77	60	84	50	24	15	30	35	40	56
41)	42	72	21	27	4	44	121	48	90	0
42)	144	60	22	48	33	0	4	88	96	84
43)	48	48	35	49	84	14	42	9	70	72
44)	0	15	100	12	10	88	77	16	60	3
45)	90	54	54	9	20	99	24	120	45	14
46)	30	45	15	24	54	144	30	18	7	21
47)	40	72	72	77	36	18	30	50	77	40
48)	55	121	44	35	36	27	108	24	35	48
49)	70	24	56	60	84	56	12	35	32	6
50)	60	0	0	9	60	132	0	99	81	36
51)	50	20	6	55	60	0	80	72	77	72
52)	12	12	64	0	72	10	15	110	27	35
53)	45	30	6	4	12	60	66	0	50	49
54)	48	35	12	21	70	80	63	63	72	11
55)	22	40	96	60	90	44	22	84	24	63
56)	24	63	0	110	132	22	110	70	24	9
57)	16	132	0	24	15	90	90	81	55	14
58)	30	0	90	18	22	60	96	16	96	110
59)	40	35	88	0	66	60	55	80	12	36
60)	28	80	88	48	63	3	48	100	44	10

© 2024 Teach Elementals, LLC

Made in United States
Troutdale, OR
11/25/2024

25206312R00064